Intermediate
Politometrics

Intermediate
Politometrics

GORDON HILTON

Columbia University Press · New York · 1976

Gordon Hilton is assistant professor of political science
at Northwestern University.

Library of Congress Cataloging in Publication Data

Hilton, Gordon.
  Intermediate politometrics.

  Includes bibliographies and index.
  1. Political statistics.   2. Regression analysis.
  3. Matrices.   4. Multivariate analysis.
I. Title
JA73.H57   320'.01'82   75–43733
ISBN 0–231–03783–X

Columbia University Press
New York   Guildford, Surrey

This book is for Tom, Agnes, and Christine

Preface

It was at dinner with Ted Robert Gurr that the idea to write this book emerged. His introductory book *Politimetrics* (*sic*) had just been published and both of us felt that another book, at a more advanced level, should be produced. But the more advanced book had to remain within the confines of the philosophy of the elementary book. This philosophy is that politometrics is a set of techniques devised to allow more theoretical flexibility—it is not an end in itself. This book is offered to political scientists with this philosophy in mind. Politometric techniques are an aid to, not a replacement for, theoretical thinking.

The book is in three parts. In the first part the student is introduced to the philosophy and formal interpretation of regression. This is done using only bivariate regression arguments, because it is essential that the student understand the assumptions underlying the regression procedure. Having grasped the fundamental arguments while discussing the most simple form of regression, the student is in a position to comprehend the more sophisticated regression procedures that appear in the second part of the book.

In the second part of the book the student is introduced to ideas of matrix algebra. Matrix algebra is useful when dealing with the multivariate regression situations encountered in Part Two and absolutely necessary for the arguments about identification in Part Three. Also in Part Two there is some consideration of regression models which are other than linear.

Part Three advances the student into multivariate multiequation situations. Political processes at their very simplest can only be accomodated by such models. Techniques are described which allow the theoretician such theoretical flexibilities as reciprocity in political processes.

To make the book as useful as possible, I have included at the end of each chapter suggested further readings of two kinds. The first is substantive and includes references to political science research which exploits some of the techniques used in the preceding chapter. In this manner I hope to provide the more pragmatic student with some reason for continuing with the book, as well as a general reading list in politometric matters. The reader will notice that as the book becomes more sophisticated, the references reduce in number. I hope that as a result of this book the next writer at this level will not have the same reference resource problem. The second type of further reading is statistical. It is included so that the student can look at the various mathematical proofs of the regression processes that I allude to but do not discuss in detail. The references in footnotes are cited in shortened form when the books or articles are listed in Further Readings at the end of the chapter.

In producing the book I have been constantly encouraged by Ted Robert Gurr here at Northwestern University. Kenneth Janda, also at Northwestern University, read an earlier draft and made some extremely valuable comments. Other than these I wish to acknowledge the tremendous help of Linda Radomski, University of Miami Medical School. She took the original first draft and carried out such a careful and responsible editing job that it was only necessary for me to rewrite parts for a final draft.

I am also very grateful to Jane Tyler of Northwestern University, who helped prepare the final draft, the copy-edited version, and the galleys. She made this normally wretched process more tolerable. Any mistakes are mine.

Evanston
June 1975   GORDON HILTON

Contents

Part One

1. Preliminaries   3
2. Understanding linear regression and correlation   29
3. Various testing sequences   60
4. Indicators of trouble in regression   80
5. Overcoming problems in regression   92

Part Two

6. Matrix algebra: visual arithmetic   117
7. Multiple regression   138
8. Various testing sequences: multivariate situations   160
9. Other regression models   186

Part Three

10. Multiequation recursive models   219
11. Structural models and the identification problem   239
12. The identification problem and parameter estimation   260

Index   281

Part One

The aim of this part of the book is to coax the student to a point where there is a basic understanding of regression. Except in odd cases all of the arguments are framed using bivariate examples. Regression models are discussed along with techniques for estimating parameters in the models. The many and varied hypothesis testing sequences are presented and clarified with examples.

The fundamental assumptions of regression are discussed at some length. Violations of the assumptions are covered as well as techniques for overcoming the difficulties presented by such violations. At the end of this section the student should have an intuitive grasp of regression. This foundation is required when the student moves into the less intuitively understandable multivariate modeling situations.

1
preliminaries

1.1 Introduction

Mathematical statisticians are aloof from the day-to-day reality of our observations of the real world. At their most useful they require elaborate experimental conditions and sophisticated measurement techniques. In both of these areas, political science is a rampant delinquent. Not only is there little scope for experimentation, but the measures and indicators we use are not precise. Political science has this in common with economics, and as economics has developed its own brand of statistics to cope with the difficulties so too can political science. While economics has econometrics, political science should develop politometrics.

The desire in political science is to be able to produce "lawlike" relationships between sets of variables. We should determine the causes and effects of political phenomena and delineate the explained and explanatory variables in our political systems. No longer can we attempt explanations or predictions with two variables only. We should drive towards multi-equation, multivariable models—models showing the interdependence between variables in the system; models capable of conforming to our intuitive notion of political dynamics, not geared to the limits of our technical skill.

In short, it is my contention that politometrics will open up new horizons for the political scientist. And this will be done in two ways. First, politometrics will generate a freedom in model building. Models will have a more realistic and flexible aspect. Instead of determining the magnitude of relationships between two variables, as we mostly do at present, we can examine what I call vector relationships. That is, we will become concerned with the direction and magnitude of a relationship rather than with magnitude alone. We will also be able to cope with some reciprocity between variables in the same system, something which we rarely see in our models but know occurs only too frequently in reality. Thus politometrics will allow for more flair and flexibility in our model construction, eventually culminating in "lawlike" relationships between sets of variables.

As yet we do not have any politometrics; all we have is the knowledge that economics is not fundamentally different from political science in data analysis. Initially we must exploit this and make econometric methods available for political scientists. We should tread the economist's path of two decades ago. But more than this, there are certain problems in political science research which require particular techniques not normally found in econometric literature. Probit analysis is an example of this. Politometrics will never really be just econometrics applied to political science. It will be a set of techniques in its own right of which a certain proportion will surely be imported from other disciplines, but a major portion will remain peculiar to the study of political phenomena. One aim of this text is to start defining the area to be called politometrics.

1.2 A politometric example

Having stated that politometrics can provide for more theoretical freedom, let us look at an example where such techniques have been used to this end. Ted Robert Gurr has been producing, over the last few years, much concerning the causes of political strife. Thus, in the two models of political strife we shall look at here, the important explained variable is *magnitude of civil strife*. In both of the models, Gurr uses various explanatory variables, various causes of the one effect—civil conflict. The choice of the explanatory variable is made on theoretical grounds, and although the second and more sophisticated model has more explanatory variables this is not due to politometric techniques. Politometrics cannot provide more variables. What they can do is provide freedom for the researcher to include the variables considered realistic to the theory.

How do econometric techniques help the theoretician? They help by providing ways of obtaining reliable estimates of the magnitude of the links between variables. Remember, estimates of the magnitude of relationships between variables can always be obtained. But only if the researcher is aware of some politometrics can we be sure that they are reliable and that theoretical inferences from the estimates be accepted with any confidence.

It is the nature of the political scientist to think causally, no matter how many protestations to the contrary there are in every research paper. It is this explicit admission in Gurr's work that leads me to use his work as an example. Gurr wants to locate the knobs and levers of society and facilitate change in society employing these knobs and levers. He constantly shows an interest in locating the explanatory variables which "cause" changes in the explained variable, although it should be noted that not all explanatory variables are necessarily manipulable.

In his progression through the studies, this is the one constant theme. But the reader should notice that the concern is with theory and better theories, not with techniques and better techniques. Because he was dissatisfied with the theoretical simplicity of a first model, he provided a means for increasing the complexity of that model to a more satisfying level. To do this he had to import considerable technical sophistication. It is important to be aware that the technical sophistication was imported to allow for more theoretical sophistication and flexibility. It was not merely brought in to play some form of technical one-upmanship.

Let us go to the first of the models. This model was reported in 1968 and concerns the relationship between relative deprivation and the magnitude of civil strife.[1] In an attempt to determine the relationship between these two variables, a series of intervening variables were suggested and diagram 1.1 shows in which way the causal structure was developed.

Notice that the ultimate explained variable is *magnitude of civil strife* which is always an effect, never a cause. Notice also that *deprivation* causes and is never an effect in this theoretical system. The other variables, *coercive potential, institutionalization, facilitation* and *legitimacy,* both cause and are caused. Considering that this is a model produced in 1968 it was relatively sophisticated for political science, which even now appears

1. T. R. Gurr, "A Causal Model of Civil Strife," *American Political Science Review* 62 (1968): 1104.

Diagram 1.1

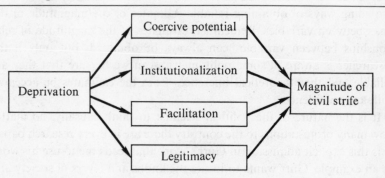

to retain this fetish for bivariate models without causal direction. Gurr is suggesting in this model that if the four intervening variables are statistically controlled for, that is, made ineffective, then any relationship between *relative deprivation* and *magnitude of civil strife* should disappear. Since the four intervening variables are the only link between deprivation and the magnitude of civil violence, if we keep these constant any change in deprivation will not influence the magnitude of civil strife. The only links to civil strife are the four intervening variables, which were rendered inoperable by the statistical procedure.

There are two salient features in this study. Notice that the model is recursive. That is, no feedback, neither direct nor indirect, occurs between any of the variables. This is intuitively naive and exists because of restrictions placed upon the causal inference technique. The restrictions derive from two technical problems. The first is a mathematical problem of identification, which is an inability to identify parameters in some reciprocal causal situations. The second problem is estimating these parameters. Thus Gurr kept away from any theoretical model with feedback.

A second feature of Gurr's preliminary analysis is the concern with correlation. This is not wrong, just limited. Although regression estimates are reported, all of the inference is made from the correlation coefficients—a common occurrence in political science. But this is really throwing out the most important part of the analysis. At its simplest form, when we regress, say, relative deprivation on magnitude of civil strife, what we are doing is developing the following linear relationship:

Civil strife = constant + weight × relative deprivation

This is "lawlike" and the most theoretically interesting feature should be the "weight" or regression coefficient. This tells us the magnitude of the influence of relative deprivation upon the dependent variable, civil strife. Now the correlation between civil strife and relative deprivation will only tell us the "goodness" of the law. It will only tell us how well the data fit. It will not tell us the "law" relating the two variables. If we move to the laws of gravity, this is rather like knowing that there is a high correlation between the velocity of a dropped object and the time elapsed since its release. But it isn't much more useful to know that the particular law is

$$\frac{\text{Velocity of}}{\text{dropped object}} = \text{constant} + 32.2 \times (\text{time since released})$$

Interest in correlation is not wrong. Correlation is just less theoretically useful than the regression coefficient.

The second Gurr model we shall look at (diagram 1.2) was produced in 1973.[2] Again the central variable is magnitude of civil strife but notice that this is not only an ultimate explained variable but an explanatory variable partially causing or influencing external intervention. Also notice the more intuitively plausible allowance for both direct and indirect feedback. This sophistication is a direct result of the author's knowledge of politometric methods. In between these two studies, Gurr wrote *Why Men Rebel*.[3] In this book he develops a sophisticated model of the causes of civil strife. The theory developed in the second paper is a precise reformulation of the theories expressed in the book. It is fair to add that the theory was only made testable and the parameters capable of estimation because of a knowledge of politometrics. As the authors report,

> Theories of the causes of civil conflict are often verbally complex. But their logical formulation and empirical evaluation have generally been primitive, at least by the standards used in constructing and testing models of macro-economic phenomena. In work published thus far, resources have been given almost entirely to testing discrete hypotheses rather than evaluating whole models. The full network of interdependencies among "independent" variables is almost never specified in advance. Reciprocal or feedback effects

2. T. R. Gurr and R. Duvall, "Civil Conflict in the 1960's: A Reciprocal Theoretical System with Parameter Estimates," *Comparative Political Studies* 6 (1973): 135.

3. T. R. Gurr, *Why Men Rebel* (Princeton, N.J.: Princeton University Press, 1970).

Diagram 1.2

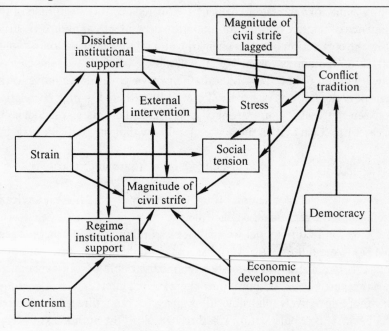

of conflict on other variables in the model are seldom specified or tested. The possibility that excluded variables systematically influence variables in the system is seldom discussed. The (usually implicit) assumptions about the technical adequacy of the system are almost never tested e.g. by examination of the error terms. Statistical techniques, particularly additive linear regression models, are applied to data without taking into account interdependencies among the "independent" variables or the basic (and vitally important) assumptions of the techniques. . . . These and other criticisms apply to our prior work as much as to others' research.[4]

As well as the excursion into reciprocal feedback, this more sophisticated model moves away from correlational analysis to regression-type analysis. Path regression estimates are reported along with the standard error of the estimates, not only allowing some idea of the size of the influence of the explanatory variable upon the explained variable but also enabling some testing of the significance of this estimate. There is thus emphasis

4. Gurr and Duvall, "Civil Conflict," p. 136.

upon the "law" relating the variables as well as the goodness of fit of the "law" to the data points. This emphasis is significant.

In summary, we see that the difference between the two models is one of sophistication. The first model is, with no allowance for any feedback, intuitively less acceptable. The second model is both agreeable in general (there will still be some researchers who cannot accept all of the relationships for substantive rather than technical reasons) and can be estimated quite accurately. An increase in technical ability has allowed a theoretical depth not possible before. As well as this, there is some focus upon the "law" relating the variables.

Finally, regression analysis also provides us with a regression line or estimate of where all the observations should be if the law were perfect. No law will be. But we can look at the observations that are away from this regression line and suggest theoretical reasons for the disparity. This may give us insight into the variables we have left out of the system (and which are causing these errors). In short, it can tell us where our model is failing, and since we can locate particular observations and examine these individually we may very well be able to reduce weakness in our models.

1.3 What is politometrics?

At the beginning of this chapter I declared that political science was a rampant delinquent in two aspects of empiricism: data condition and experimentation. The condition of political science data leaves much to be desired, but this is something that can be remedied. There is no fundamental reason why the data cannot be improved. The condition of our data will be some function of the energy we put into their collection.

However, this is not so of experimentation. Political science inherently does not lend itself to experimentation, and this nonexperimental nature is really not changeable. We have to be content with examining how man behaves in the real political world. We also have to cope with the difficulties this produces for us when we try to insert such empirical content into our theories. Politometrics is an acknowledgment of these difficulties and provides guidance in how we might get around them.

Earlier I said that mathematical statistics were not really relevant to the real world; this is so. As science moves towards the real world, so more and more it violates the assumptions built into statistical theory. As we progress into empirical situations, we begin to swerve from the purity that characterizes the experimental method. In violating these assumptions we place ourselves in danger of making erroneous inferences from our data. Politometrics allow this violation of theoretical statistics by providing us with techniques which cope with the violations. Politometrics will furnish statistical maneuvers which reduce the possibility of incorrect inferences.

Let me see whether I can make this more clear. So that political science can exploit the power of mathematical formulation with its capacity for manipulation and deduction it has to generate parametric models. Obviously, theories, laws, propositions are concerned with relationships between concepts or their surrogate variables. Consider the general law, $Y =$ function of X, where "function" represents some mathematical relationship. In its most simple form the above relationship can be rendered more specific by allowing a linear function. This would produce

$$Y = a + bX \quad \text{where} \quad \begin{aligned} a &= \text{constant} \\ b &= \text{weight component} \end{aligned}$$

This model states that any unit change in X produces a bX change in Y. Thus, the model is arranged to provide that Y is caused by X.

Why then is it important to know the values of the parameters a and b? There are two reasons. First, if we know both of the parameters and have some value for X, we can predict the value of Y. So there is help with prediction. Second, it has theoretical significance. If, in some manner, we can show that b is equal to zero, what we are doing is proposing that Y is unaffected by X. The importance of this must be evident. We can determine, in this manner, which variables belong in any model and which do not. We need to develop such models if we are to move toward the knowledge power obtained in the physical and biological sciences.

Having developed, for theoretical reasons, simple models, we now need to provide some facility for obtaining empirical estimators of the parameters. Regression analysis is the most common. We regress X on Y using observations of X and Y, and together with some criteria for calculation, we can obtain estimates of a and b.

Initially regression was devised to compare the height of fathers with their sons. Notice this is a nonexperimental situation. It's ironic that regression began in nonexperimental situations but was immediately imported into experimental situation where the various assumptions of regression analysis were developed. Now we have to relax these assumptions to facilitate the use of regression in nonexperimental situations again.

The basic assumptions of regression analysis were determined in the following type of experimental situation. Let us suppose there is interest in the effects of various quantities of fertilizer on the growth of some plant. To determine the effect of the explanatory variable (the fertilizer) on the explained variable (the plant) we would insert the seeds of the plant into plots with various mixtures of the fertilizer and soil. Throughout the growth of the plant we would be careful to water each plot exactly alike and provide equal amounts of sunlight. In this way we have manipulated the explanatory variable so that we can measure its effect upon the explained variable. Notice there is no randomness in the explanatory variable. One plot has X_1 pounds of fertilizer per square yard and other plots have X_2, X_3, X_4, and so on. And assuming we can measure accurately, these will be precise amounts.

In political science we want similar models. But there are no situations where we can manipulate the explanatory variable in the above fashion. If we take them as they are, uncontrolled, we are deviating from the assumptions of regression as it was developed in the biological sciences and may be in danger of making erroneous inferences. Politometrics guards against this, to a large extent, allowing us to violate these assumptions and still retain a facility for estimating the parameters we are interested in. Politometrics is a set of techniques for coping with real-life empirical situations when experiments are not possible.

In summary, political science deals with relationships between political variables. The most powerful models of these relationships are parametric. The objective of politometrics is to give empirical content to these parametric models. There is, thus, a concern for statistical inference from the data back to the models. Politometrics is not used to develop theories. It is used to estimate parameters of an already existing theory and as such is a step in the process of inquiry. Politometrics differs from theoretical statistics in that it can cope more easily with the real empirical situations that confront the political scientist. The politometrician therefore (1) takes theories of political behavior and provides measures of the relationships therein, and (2) uses empirical parameters to predict events in the political sphere.

In the dichotomy of normative political science and positive political science, where the former is concerned with value judgments and concepts of moral obligation and the latter with propositions which are questions of fact, politometrics is almost exclusively positive political science. If we are concerned with the effect of relative deprivation on the incidence of civil strife, the positive question is to be able to forecast a change in the incidence of civil strife by some change in relative deprivation; the normative question might be whether such effects are moral.

1.4 Populations and samples

A population, as used in statistics, contains every conceivable observation of some specified variable. For instance, all the riots in the United States would be one population; all the riots in the United States in 1968 would be another population; all the riots in the state of Alabama in 1968 would be a further population. Notice that the second and third examples would be included in the first population and that the third population would be found within the second. That they are subsets of one another does not deny them the possibility of being populations. Whether some set of observations is a population or not depends upon the use of those data.

There are some situations where it is not possible to collect observations on the whole population. For example, we might not know the bounds of the population. It could be the case that some of the theoretical observations in the population have not occurred. The incidence of heads or tails in coin tossing is presumably of this nature. Thus we see that populations can be infinite.

Even if a population has finite characteristics it may not be possible to collect all observations within the population. Financial resources, or lack of them, may prohibit this.

In both of the above situations we end up with a sample of the population. Thus a statistical sample is made up of a limited number of observations. The whole point of statistics is to allow us to make inferences about the population using the information obtained from the sample.

Statistics provides for the generation of estimates of various summarizing measures for whole populations using limited samples from those populations. Accompanying these measures will be other measures indicating the unreliability of the estimates obtained from the sample. Let us take an example. Suppose we are looking at two variables; the first is the salary of an individual Republican and the second is the amount of money donated by an individual Republican to Richard Nixon's 1972 Presidential campaign. Suppose our population is all the Republicans in the United States. Further, let us suggest that there is some "law" relating these two variables and this "law" provides that the higher the salary the higher the contribution. Thus we can regress salary on contribution and the regression coefficient will show how much in campaign contributions can be expected from someone with a particular salary. Let us call this population parameter β. Obviously this population is finite and would include every Republican who draws a salary in the United States. Pictorially our operation could be represented in diagram 1.3. The regression model for this would be

$$Y = \text{constant} + \beta X$$

Diagram 1.3

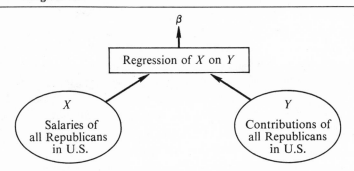

Unfortunately, however, we may not be able to sample every Republican in the United States. We may only be able to sample 10 percent. Thus we randomly sample this proportion of the population, producing a sample of Y and a sample of X. Regressing our sample X on our sample Y, we obtain the sample parameter $\hat{\beta}$ (diagram 1.4). Given that we have developed the estimator $\hat{\beta}$, what can we say about $\hat{\beta}$? This, of course, depends upon the characteristics of our estimator.

Diagram 1.4

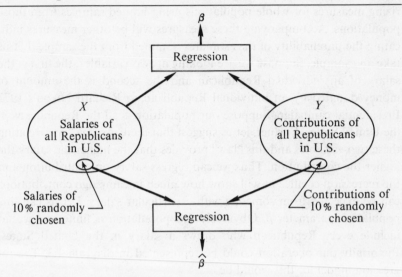

1.5   Characteristics of estimators

We can take more than one sample. Indeed it may be possible to take an infinite number of them. As diagram 1.5 shows, we will produce, if we regress each sample X on sample Y, as many estimators of the population parameter as we have samples from each population.

Because none of the samples will be the same, we can expect to get a distribution of estimators. If the population that the samples come from is normal then the distribution of the estimators will be normal. This sampling distribution of estimators will have a mean and variance. We can call the mean of the sampling distribution the expected value of the distribution.

The variance of the sampling distribution will be related to the sample size. For example, if we have a population of 100 observations and take a series of samples of five we would expect to get a sampling distribution with a large variance. The smallness of the sample will facilitate the occurrence of unrepresentative samples. Thus very deviant estimators will occur more frequently in small samples than in large samples. Generally, we can say that as the sample size increases, so the variance of the sampling distribution decreases (diagram 1.6); an important property, as we shall later see.

Diagram 1.5

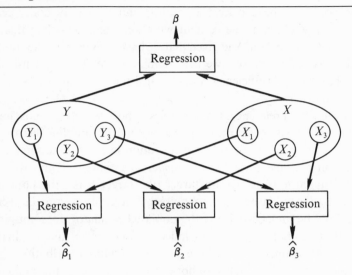

Diagram 1.6

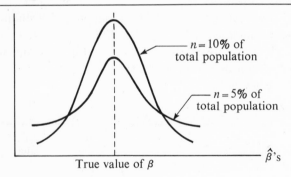

True value of β

The process of inferring from a particular sample to its parent population is a function of four components: (1) the population, (2) the relationship between this population and the many samples taken from the population, (3) the relationship between the characteristics of the many samples taken from the population and the characteristics of the one sample, and (4) the characteristics of the sample we observe. It is these four components which determine the quality of our estimator.

The perfect estimator would always give us the true value of the population parameter β. But perfect estimators rarely exist. While dipping one's toes into a swimming pool to get some estimate of the temperature of the pool may be a perfect estimator (and then only under certain conditions), we have to be content with something less than this with most social science data. And given that we have to determine the goodness or otherwise of estimators, we need some criteria. The most useful are (1) unbiasedness, (2) efficiency, and (3) consistency. I shall deal with each of these in order.

In dealing with references from samples to populations the most important characteristics are the mean and variance of the sampling distribution. When the mean of the sampling distribution is exactly the same as the population parameter we have an *unbiased* estimator. Thus, while the perfect estimator is always on target (the target being the population parameter), the unbiased estimator gives the correct answer on the average.

It should be understood that the unbiased estimator does not give us the perfect estimate every time—indeed it may never give us this! All unbiasedness claims is that in the normal sampling distribution half of the sample estimators will be above the true value of the population parameter and half below.

More formally, if we define the mean of the sampling distribution as $E(\hat{\beta})$ then we have an unbiased estimator when $E(\hat{\beta}) = \beta$. Thus, bias can be defined by the simple transformation, bias $= E(\hat{\beta}) - \beta$. And when bias is zero this relationship reduces to $E(\hat{\beta}) - \beta = 0$. In frequency distribution form we get diagram 1.7.

Diagram 1.7

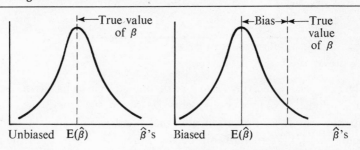

In making inferences, an unbiased estimator is of little comfort if the dispersion or variance of the estimates in the sampling distribution is very large. In this case the standard error of the estimator, that is, the square root of the variance for the distribution, will be large, and although it's nice to know that the average of the estimators is equal to the true value of the population parameter, there is only a small chance that any particular estimate is close to the true value. For instance, suppose that the true value of β is zero. Our unbiased criterion says that the sum of all the deviations of all the estimators from the true value is equal to zero. Thus, if the true value of β is zero, a sampling distribution with values of $+50$ and -50 is as unbiased as one that has values of $+\frac{1}{2}$ and $-\frac{1}{2}$. The one with the majority of its values closer to the mean of the sampling distribution and thus the target value is better because any one estimator is bound to be pretty close to the target value. In diagram 1.8, sampling distribution A is far better than sampling distribution B although both are unbiased.

Diagram 1.8

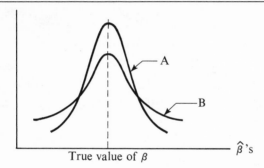

The standard error of the estimator of sampling distribution A is smaller than that of B. This gives us our clue to what we mean by *efficiency*. An efficient estimator is one which comes from a sampling distribution with the smallest dispersion.

However, this definition of efficiency is relative because we will never know the sampling distributions of all possible estimators. We can relax this definition to obtain *relative efficiency* which occurs when any estimator comes from a sampling distribution whose variance of the estimator is less than that of any other known sampling distribution. More formally, $\hat{\beta}$ is an efficient estimator of β, if the following conditions are satisfied:

    (1) $\hat{\beta}$ is unbiased,
    (2) variance $(\hat{\beta}) \leqslant$ variance of any other estimator of β.

It should be noted that even the most slightly biased estimator cannot be called efficient if it has the smallest dispersion in its sampling distribution. Notice also that there may very well be a need to compromise between bias and small dispersion. Consider diagram 1.9 with sampling distributions C and D.

Diagram 1.9

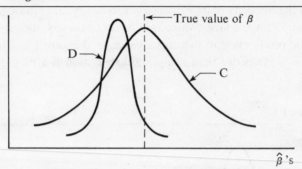

Distribution C is unbiased but has very large dispersion. Distribution D is a biased estimator but has a small dispersion of values around its mean. A larger proportion of the estimators from sampling distribution D are closer to the true value of β than those from distribution C. We cannot say which of the two distributions is better unless we are inclined to place relative importance to unbiasedness and small dispersion.

Our third criterion, *consistency*, is concerned with the behavior of an estimator when the sample size is increased. A consistent estimator is one which centers closer and closer to the target population parameter as the sample size increases. In a sample as large as the population itself, the estimator will be the true value of the population parameter. Consistency is an admirable quality for an estimator to possess because it means that we can buy increased reliability in our estimating by increasing the sample size. With an inconsistent estimator it would not really matter whether we had a sample of 1 or 100. Diagram 1.10 shows what happens to a consistent estimator as sample size is increased. As the diagram shows, the consistent estimator need not necessarily be unbiased nor have the smallest dispersion in its sampling distribution. For example, we normally use sample standard deviation of a distribution, s, as an estimator of the population standard deviation, σ. But when the sample size is small, s is

Diagram 1.10

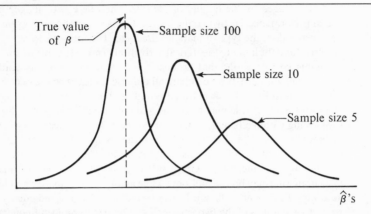

a biased estimator of σ and we have to use a Student t-distribution to compensate for the bias. But as the sample size gets larger, so Student's t-distribution approximates the normal distribution and we can conclude that s becomes a good estimator of σ as the sample size increases. Thus s is a consistent estimator of σ.

If only in increasing the sample size does the bias reduce to zero, then we can say that the estimator is asymptotically unbiased, where asymptotic means large samples. The criteria we have discussed here are all for small-sample situations, this being the most common research situation in political science. These criteria are also the most difficult to satisfy, and it can be said that if an estimator satisfies these small-sample criteria then it will also satisfy the large-sample criteria. For a good discussion of large-sample or asymptotic criteria see Goldberger.[5]

Let me use just one more practical example to compare and contrast the three criteria. The example comes from Kmenta.[6]

> Estimating a parameter from a sample can be compared to our shooting at a target with a rifle. In this parallel, the bull's-eye represents the true value of the parameter, each shot represents a particular estimate (sample), the rifle is our estimator (i.e. estimation formula), and the distance from the target reflect our sample size. In reality we normally have only one sample and thus can make only one estimate; that is—in our parallel—we are

5. A. S. Goldberger, *Econometric Theory*, pp. 127–30.

6. Jan Kmenta, *Elements of Econometrics*, p. 13.

allowed only one shot. However the quality of any shot before it is made is clearly judged by the quality of the rifle. The rifle can be judged by its actual performance, i.e. by making a large number of shots, or by examining its construction, the type of material used, etc. The former corresponds to empirical and the latter to theoretical derivation of properties of an estimator. An unbiased rifle is one that produces shots that are randomly scattered around the bull's-eye. If we compare all unbiased rifles, then the ones whose shots are most heavily concentrated round the bull's-eye can be considered efficient. Finally, a rifle may be considered consistent if the probability of a shot falling within a (small) distance from the bull's-eye increases when the distance between the shooter and the target is decreased. Note that the quality of the rifle is judged by its repeated performance (actual or expected) and not by the single shot. Given just one shot, it may happen that an inaccurate rifle may hit the bull's-eye while an obviously superior and highly accurate rifle may not. Obviously, this would not affect our judgment of the respective qualities of the two rifles unless it tended to happen frequently.

Finally, in this section let me discuss the most desirable of all estimators, the BLUE estimator. This is an abbreviation for *Best Linear Unbiased Efficient* estimator. In the universe of all estimators there will be those that are linear, that is, the estimator constructed is a linear function of the observations. Two examples of this are the mean of a single distribution, and the regression coefficient relating two distributions. In this reduced universe of linear estimators, there will be a subset which contains all the unbiased estimators. And within this further reduced universe of linear unbiased estimators there will be one estimator which had the smallest dispersion of its values around the target parameter. This estimator is BLUE. More succinctly, the BLUE estimator

(1) is constructed as a linear function of the observations,
(2) is unbiased,
(3) has a variance which is smaller than the variance of any other estimator.

1.6   Hypothesis testing and statistical significance

It is not my intention in a book at this level to explain the various theoretical foundations to hypothesis testing and statistical significance. Nevertheless, if we are to be using politometrics to aid in theory testing, it would be wise to consider some of the less obvious features of these procedures. Having considered various estimators and the sampling distributions from which they come, we will want to test various hypotheses about the estimators. For instance, the most basic hypothesis is concerned with whether an estimator is significantly different from zero. In the regression equation $Y = \alpha + \beta X$, if β is not different from zero the equation reduces to $Y = \alpha$. We can conclude that X does not influence Y—an important theoretical conclusion.

The statistical decision process consists of:

(1) choice of the significance level,
(2) determination of the acceptance/rejection region,
(3) selection of the test statistic,
(4) the decision.

We can use these steps as headings to order some relevant points.

1.6.1 Choosing the significance level

Almost exclusively the 5 percent significance level is selected. Arguments for this decision include those concerning consistency in research across the profession. In this way scientists have some idea of the comparability of various research results. But this significance level is arbitrary and based mostly upon the conventions of journals. Such rigid adherence to one level of significance with its implications of what is, and what is not, considered an interesting result may have political consequences in the social sciences in general and political science in particular, especially in situations where political practitioners and their aides have access to the research.

Let me review briefly the various errors that one might make in testing hypotheses. There are two types of error: Type I error is rejecting a true hypothesis; Type II error is failing to reject a false hypothesis (see diagram 1.11). These errors are inversely related to one another. Thus, as you take steps to decrease the probability of making one error, you necessarily increase the chance of making the other. Because concern with the dangers associated with making these errors is onerous, most of us find relief in accepting the journal-given holy 5 percent level. But there is increasing dissatisfaction with this policy, and there is now a school of statisticians who advocate that the researcher should look to the substantive consequences of making either a Type I or Type II error when selecting a significance level.

Diagram 1.11

		Hypothesis	
		True	False
Decision	Accept	Correct	Type II error
	Reject	Type I error	Correct

An example might be appropriate at this point. Let us take research into the Headstart program for educating the most underprivileged children of America. In trying to determine the effectiveness of the program, one might take children who have been involved in Headstart and compare them with children who have not. Assuming that we have a controlled experiment to take out all the other relevant variables, our null-hypothesis might be:

H_0: There will be no difference in educational level between individuals who have been in the Headstart program and those individuals who have not.

Having set up this hypothesis, we now require some significance level. If we choose a low significance level, say, 0.01, there will be more chance of accepting the null-hypothesis, whether it be true or false. If it is a true null-hypothesis then no damage has been done; however, if it is false we have committed a Type II error. By committing this kind of error we have increased the evidence for those who want to kill the Headstart program. In making it very difficult to reject this null-hypothesis one has, perhaps unwittingly, taken a political position on Headstart.

Suppose, however, that the researcher chooses a higher significance level, say 0.30. The possibility of rejecting the null-hypothesis is considerably increased. If we do reject the null-hypothesis and it is a true hypothesis, then we have made a Type I error. But what has also happened is that we have provided evidence for those who want to retain the program. Again we have taken a political position.

There has been a revulsion on the part of scientists to taking such positions. But it is not a novel radical posture; indeed, so-called objective scientists have been taking political positions by default in choosing the

5 percent level of significance automatically. The political position should be chosen deliberately and with awareness.

The significance level can be chosen to take into account the dangers, both political and substantive, of increasing or decreasing the probabilities of making the two types of error. What we are doing is bringing into control certain aspects of our scientific decision which hitherto have been out of control. This must be more scientific. Different kinds of research problems should command different types of decisions concerning levels of acceptance and rejection. This would be an effort to relate the real-life consequences of our research to the scientific decisions we make.

There are some who take a much more radical view on significance testing and prefer to leave the readers to select the significance level they find relevant. Thus they would report only the information which allows the scientist to obtain the significance of the result, in terms that are significant to him. This also, it is argued, prevents the layman practitioner from being misled by the magic words "statistically significant."

1.6.2   Determination of the acceptance/rejection regions

Having chosen a particular level of significance, for whatever reason, we are in a position to select the regions of acceptance and rejection. Three patterns vie for attention here. We can either have one-tailed tests, two-tailed tests (symmetrical), or two-tailed tests (asymmetrical). Again I shall argue for relating one's selection to the practical research situation rather than accepting the convention blindly.

It is a rare research situation where the researcher cannot predict the sign (and thus the direction) of the test statistic. This being the case, calling for a two-tailed test is pointless. If we know that the testing area will be in one tail of the distribution only, our selection should reflect this. Obviously, if the researcher is really completely uncertain as to the direction of the hypothesis then a two-tailed symmetric test is appropriate.

The asymmetrical two-tailed test presents other complications. When we are really doing is taking our already chosen significance level and dividing the percentage unequally between the two tails. Diagram 1.12 shows this. The way in which we divide up the original area of rejection and thus acceptance will give us varying probabilities in either tail of the distribution of making either Type I or Type II errors. This, as before, will have political implications. Our research decision as to the asymmetry of the two tails should be scrutinized for all possible consequences.

Diagram 1.12

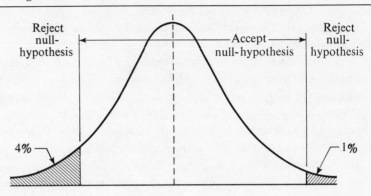

1.6.3 Selection of the test statistic

In most research situations there is little choice among test statistics; they are normally thrust upon the researcher. Nevertheless one can relate the desirable properties of the estimator, those we discussed earlier, to the whole testing procedure by considering the test statistic.

Most test statistics are of the form

$$\frac{\text{estimator} - \text{hypothesized population parameter}}{\text{standard deviation of the estimator}}$$

In the regression situation this becomes

$$\frac{\hat{\beta} - \beta}{\text{standard error of } \hat{\beta}}$$

Obviously, the size of the test statistic is influenced both by the estimator and the standard error of the estimator. Given any hypothesized population parameter, the absolute value of the test statistic will be a function of both. Thus a *biased* estimator will clearly produce an incorrect numerator, which may result in an incorrect inference from the sample. For example, if our null-hypothesis is $\beta = 0$ and we obtain a biased estimate of this parameter equal to 2.0, this will increase our chances of rejecting the null-hypothesis incorrectly, a Type I error. Having a biased estimator may also result in turning a symmetrical two-tailed test into an asymmetric two-tailed test.

The second desirable property of the estimator is *efficiency*. We said that the most efficient estimator is the one where the variance of the sampling distribution is smallest. Let us see what this means in terms of acceptance or rejection of our hypothesis. The smaller this standard deviation, as diagram 1.13 shows, the less will be the acceptance region for the null hypothesis. Note that we do not select the most efficient estimator in order to obtain an acceptance or rejection region to our liking—the smallest variance is desired because it says something about the quality of our summarizing measure, the estimator—but it will have some consequence for our acceptance and rejection decisions.

Diagram 1.13

Both distributions at 5% two-tailed symmetrical acceptance regions

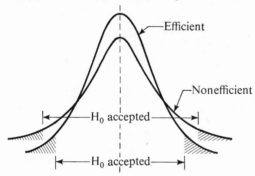

Consistency was our third desirable quality of the estimator and concerns the behavior of the estimator as the sample size is increased. The estimator is required to become less biased and more efficient as sample size increases. For the reasons stated above, a consistent estimator will have some consequences for the test statistic.

1.6.4 The decision
The research decision will be a function of the three previous steps in the decision process and the observations we operate with. But this is not what I am concerned with here. Rather, I am more interested in the dangers, for the social sciences, of letting data be the only criterion upon which one determines the acceptance or rejection of theory. Following the tradition of the natural sciences, if there is confrontation between the data and the theory, then the theory is declared lacking. Someone once described this as the "one-way bulldozing of theory by data." Considering the sloppiness of most social science data, this is not a wise policy. Consider

diagram 1.14. Let us take the most simple grades of data: those which are good and those which are bad. Let us consider both true and false hypotheses. In obtaining a rejection of our hypothesis, which of the interacting situations is it? The one where we have good data and a false hypothesis or the condition where we have bad data and a true hypothesis? This type of situation is allowed for in the statistical testing, it is claimed. But is it? Most testing procedures were developed for relatively precise data. Until the social sciences have such data we are not wise to allow this "one-way bulldozing."

Diagram 1.14

		Theory	
		True	False
Good		Accept	Reject
Bad		Reject	Accept

Data

All the ideas in this section are presented as thought-provokers. Any reader of this book should be well past the stage of having such testing procedures described—this should have been done in earlier, more elementary statistics courses. My wish here has been to present relevant but frequently overlooked consequences of the classical inference-testing processes.

1.7   The problem of hypothetical populations

Inference testing, as we have seen, deals with obtaining knowledge about populations using information from samples. This is done in compliance with carefully defined sets of procedures. But what do we do in the common political science situation where the sample equals the population? For example, if we collect, for all nations, observations on some variable for a particular year, do we not have the whole population there? And, if we have, what then would be the meaning of any significance testing that we carry out? Having carried out statistical significance testing, what precisely is the universe about which we are making inferences? In the experimental sciences, the researcher is reasonably confident that, should the experiment be repeated, another set of observations would be obtained—another sample. And presumably, financial resources allowing, one could obtain many samples. So any inference is made from one of these experimental samples to some presumably infinite population.

In the nonexperimental sciences, observations cannot be repeated. Political behavior exists at a particular time under certain conditions which may never occur again. Thus, the sample we have constitutes the population. Nevertheless, one can imagine, if not actually produce, a hypothetical population. In any situation there will be certain major influential conditions. At the same time, operating in each particular situation there will be random chance factors. These random chance factors can produce a hypothetical population of observations given the basic determinants of the situation. So in the one-off situation that we observe, the random chance factors produce the set of observations that we get. But at the same time, the chance factors could have produced a potentially infinite distribution of samples. Given a chance to repeat the situation—the "experiment"—with the identical major influential conditions as the first occurrence, would we expect the same observations? The answer has to be no, since the random chance factors would generate a different set of observations. But these observations would not be too unlike the original observations since the major elements of the two situations were the same. One would expect minor variations in the observations from one to the other. But is this not what the experimentalist gets?

Thus when we carry out significance testing upon samples which apparently seem the same as their parent population, our inference is from that sample to a hypothetical population, irreproducible but certainly imaginable.

Further readings

Substantive

Bowers, W. J., and R. G. Salem. 1972. "Severity of Formal Sanctions as a Repressive Response to Deviant Behavior." *Law and Society Review* 6: 427.

Cameron, D. R., J. S. Hendricks, and R. I. Hofferbert. 1972. "Urbanization, Social Structure and Mass Politics: A Comparison within Five Nations." *Comparative Political Studies* 5: 259.

Cnudde, C., and D. J. McCrone. 1966. "The Linkage between Constituency Attitudes and Congressional Voting Behavior: A Causal Model." *Am. Pol. Sci. Review* 60: 66.

Davis, O. T., M. A. H. Dempster, and A. Wildavsky. 1966. "A Theory of the Budgetary Process." *Am. Pol. Sci. Review* 60: 529.

Goldberg, A. S. 1966. "Discerning a Causal Pattern among Data on Voting Behavior." *Am. Pol. Sci. Review* 60: 913.

Jacobson, A. L. 1973. "Intrasocietal Conflict: A Preliminary Test of a Structural-Level Theory." *Comparative Political Studies* 6: 62.

Miller, W. E., and D. E. Stokes. "Constituency Influence in Congress." *Am. Pol. Sci. Review* 57: 45.

Muller, E. N. 1970. "Cross-National Dimensions of Political Competence." *Am. Pol. Sci. Review* 64: 792.

Seitz, S. T. 1972. "Firearms, Homicides, and Gun Control Effectiveness." *Law and Society Review* 6: 595.

Sharkansky, I. 1968. "Agency Requests, Gubernatorial Support and Budget Success in State Legislatures." *Am. Pol. Sci. Review* 62: 1220.

Winham, G. R. 1970. "Political Development and Lerner's Theory: Further Test of a Causal Model." *Am. Pol. Sci. Review* 64: 810.

Statistical

Throughout the book the further readings in statistics will be taken almost exclusively from the four books listed below. In later chapters, they will be referred to by the authors' surnames. They are, in majority opinion, the best books at this level. Obviously, many of the arguments proceed with economic variables, but the student should still try to go through the arguments. Nothing is easy at this level of politometric sophistication. The books are in alphabetic order.

Goldberger, A. S. 1964. *Econometric Theory*. New York: Wiley, pp. 1–4.

Johnston, J. 1972. *Econometric Methods*. 2d ed. New York: McGraw-Hill, pp. 1–7.

Kmenta, J. 1971. *Elements of Econometrics*. New York: Macmillan, pp. 1–17, 197–201.

Wonnacott, R. J., and T. H. Wonnacott.1970. *Econometrics*. New York: Wiley. No particular reading for this chapter.

2

understanding linear regression and correlation

An article of faith for each scientist is regularity in the behavior of the various concepts of his interest. A reality in which chosen concepts behave randomly, as if unrelated, does not facilitate theoretical speculation. Our empirical problem is to determine the hidden regularities in our data. And from a theoretical perspective we are concerned with relationship between concepts. If these relationships can be shown to remain regular, then confidence in our theory is increased. Thus, we desire to develop models which portray our theoretical beliefs. Having developed such models, we need to estimate the various parameters which link the variables in the model.

2.1   Deterministic and stochastic models

An equation has two sides divided by an equal sign. In the models we shall be dealing with, on the left-hand side of the equation will be a single variable. This is variously called the dependent variable, the effect variable, and the regressand, but we shall refer to it as the *explained variable*. The *explained variable* is explained by those variables on the right-hand side of the equation. These also have varying names such as cause variables, regressors, or independent variables, but since they are rarely independent we shall refer to them as *explanatory variables*. Thus all models consist of an explained variable on the left-hand side of the equation and one or more explanatory variables on the right-hand side.

We will discuss two types of models. The *deterministic* model is one in which the explained variable is perfectly explained by the explanatory variable. This means that all of the causes of the explained variables are to be found on the right-hand side of the equation. Given that we can obtain a set of values for these explanatory variables, then we can predict accurately the value of the explained variable. A deterministic model exists when we have perfect knowledge in our research situation.

For example, suppose we have the model, $y = 2x$. Then for every observation of x we will obtain a specific value of y. And all equal observations of x will produce equal values of y. If we plot observations of x and y on a scattergram, we will obtain a line upon which every pair of observations of x and y is located (diagram 2.1). Although the example given is linear, nonlinear deterministic models also exist. In these situations the lines will be curvilinear but still with all the observation points located exactly on the line.

Diagram 2.1

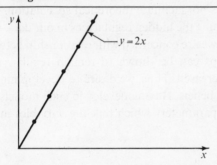

It is rare, however, to find oneself in this glorious position of full knowledge. Variables which are explanatory but omitted from the model represent our lack of knowledge about the phenomenon under study. There is an unexplained component in our models: a portion of behavior of the explained variable which we are not explaining. This is a *stochastic model*. It is an admittance of imperfection in our theoretical knowledge. In modeling this type of situation we have to add to the known explanatory variable a further term in the equation which represents all of the explanatory variables that we have left out of the model. This term is called the error term and its effect is to disturb the data points away from the line relating the explained to the explanatory variables.

The error term can also embrace two other kinds of weakness in the total research procedure. The first concerns difficulties in the research method and involves slippage between what we attempt and what we achieve. The other is completely out of the hands of the researcher and concerns the purely random element brought into every research situation and phenomenon. There is little to say about this except to recognize its existence.

But the slippage in the research method does warrant discussion. There are two research steps of which we cannot be sure. First, we do not know whether all of our measurements of the explained variable are accurate. Any error in our measurement of the explained variable may be allowed for within the error term of the regression model. In political science, with the admitted difficulties in measurement, the measurement error might be considered a large portion of any error term inserted in the regression model.

Second, even if we are sure of the measurement of the explained variable (remember, so far we have always insisted on perfect measurement of the explanatory variables) and have specified our model correctly, we may still not be able to interpret all of this in a correct mathematical model. For example, although we are sure that the concept of military power can be usefully inserted in many of our political science models, and that we can measure the numbers of tanks, bombers, ICBM's, etc., very accurately, we are a little hesitant not to include an error term which would allow for the possibility that this measure, although accurate, may not really indicate what we want it to. We may also import error into our equations by making the simplifying assumption of linearity in situations where it is not really appropriate.

The error term in the regression equation is thus included as a catch-all for:

(1) pure random error;
(2) misspecification—explanatory variables omission
            —incorrect mathematical formulation;
(3) research difficulty—measurement error.

Let us return to our model but now add a general mop-up error term. In the model $y = 2x + u$, where u is the error term, u can take on any value, positive or negative. When we plot this on a scattergram we might obtain the pattern shown in diagram 2.2.

Diagram 2.2

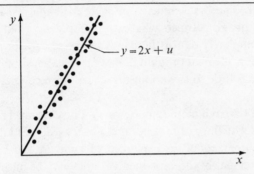

$y = 2x + u$

Notice in this scattergram how the data points are away from the line. Obviously, the farther away they are from the line the larger the error component in the model. It is fair to say that most political science models leave more unexplained than explained. Indeed, no social science model should ever be deterministic, since a researcher must balance explanatory power with parsimony. A model that explains 90 percent of the variance in some explained variable using five explanatory variables may very well be more efficacious than another model with 50 explanatory variables and an explanation of 95 percent of the variance. Finally, since I know of no deterministic model in political science, we shall be dealing exclusively with stochastic models.

2.2   Regression models: an intuitive discussion

We ultimately want to express relationships between variables in some mathematical form. In the most simple case, the linear form, we want to estimate the parameters in the regression model:

$y = \alpha + \beta x + u$ where α and β are parameters, y is the   (2.1)
variable to be explained, x is the
explanatory variable, and u represents
all the explanatory variables left out of
the model

To obtain α and β we need to regress x on y. An example might clarify. Let us be interested in the relationship between the amount of money spent in a particular ward on a campaign and the percentage increase in votes for the party spending the money. We are using amount of money spent as an explanatory variable to determine increase in percentage of the vote.

y = increase in percentage of the vote

x = amount of money spent in the ward

In this experimental adventure, we give \$1000 to three wards, \$2000 to three other wards, \$3000 to a further three wards, and so on up to 18 wards. For each of these wards we obtain the increases in the percentage of the vote for our party and we plot all of this on a scattergram, as in diagram 2.3.

Diagram 2.3

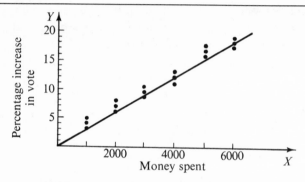

As our model is of a straight line and there is no one straight line that will go through all of these points, we have to devise some method by which we can produce the straight line which best represents the observations. The points off the line will give us some assessment of the goodness of our model. If we had a fitted straight line which was some distance from many of the observations, it would indicate an inappropriateness of our model. We could, in this situation, conclude any or all of three things: First, the relationship is not linear and thus our model is incorrectly specified. Second, the model is linear, but the variables left out of the model are extremely influential on the explained variable. Third, our method for measuring the explained variable is too imprecise.

Looking at our example, there is certainly some linear relationship between our two variables. As the money spent increases so does the increase in percentage of the vote. One can picture some linear band

Diagram 2.4

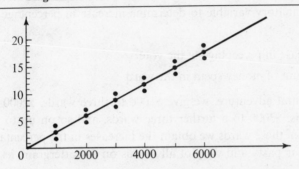

which covers all of the points. Our task is to reduce this band to a line. Let us visually place a line through our points (diagram 2.4). Notice that for each amount of money spent there is variation in the resulting increase of percentage of the vote. This variation will perhaps be caused by the unknown variables that we have left out of the model and which are represented by the catch-all error component, u. Also notice that this "error" causes observations to fall on either side of our visually placed regression line, since the unknown excluded explanatory variables can cause positive or negative errors.

What then are the parameters α and β? Let us look at α, the regression constant, first. If we spend zero money in a particular ward, our line would predict that there is no increase in percentage of the vote. In this case, our regression model would state that when x is zero (no money is spent) y will be equal to zero (there will be no increase in percentage of the vote). If we return to our regression model, we obtain the relationship

$$0 = \alpha + \beta \times 0 + u$$

As we shall see more fully later, the regression line is located so that the average value of the error term u will be zero. The above equation can therefore be reduced to

$$\alpha = 0$$

The value of α gives the point at which regression line cuts the axis of the explained variable. And α can take any value from minus to plus infinity.

The parameter β represents the slope of the regression line and is some measure of the rate at which a change in the explanatory variable produces a change in the explained variable. Let us consider the three slopes shown

Diagram 2.5

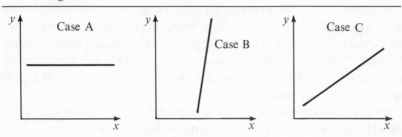

in diagram 2.5. In case A the regression line is horizontal, and thus the slope of the line is zero. Substantively this means that any change in the explanatory variable produces no change in the explained variable. Thus, there is no relationship between our explained and explanatory variable. In our regression model when $\beta = 0$ the model reduces to

$$y = \alpha + 0 + u$$

which prescribes that y is dependent upon α and u, neither of which is a known explanatory variable.

In case B the slope of the line is almost vertical. This implies that an increase in the explanatory variable produces a tremendous increase in the explained variable. The explained variable is extremely sensitive to changes in the explanatory variable. The slope in this situation takes on a value approaching infinity. Were the slope equal to infinity, the regression line would be perfectly vertical. This rarely happens. Much more likely is the slope seen in case C. In this situation any change in the explanatory variable has a modest effect upon the explained variable, but nevertheless a substantively significant effect. It is the responsibility of regression to exploit data to determine whether such effects exist.

The discussion of the regression slope, thus far, only allows for values of β which range between zero and infinity. However, the regression slope can also range from zero to minus infinity. In such cases the substantive implication is that as the explanatory variable *increases* the explained variable *decreases*.

Having located a line (we shall discuss how in great detail shortly), how do we calculate the value of β? Simply, in the same manner that we would employ to obtain the slope of any gradient. We calculate how far we go up on the vertical scale as we move along one unit on the horizontal scale. Returning to our example, we see that as we move between $2000 and $3000, that is, an increase of $1000, we move on the vertical scale from 5 percent to about 8 percent in improving the voting for our party; an increase of 3 percent. Thus $1000 gets us 3 percent and our parameter becomes 3/1000, units being percentage increase per dollar spent. Notice that the slope coefficient is dependent upon the units chosen. We shall see later that other popular measures are unitless.

Finally, in this first skirmish with the ideas behind regression, the student should understand that although the discussion has been in terms of two variables there is no reason why we cannot generalize these ideas to more variables. If we have three variables, one explained and two explanatory, then our regression is in three-dimensional space and the regression line is promoted to a regression surface. Each additional variable that is added changes the fundamental character of this surface. And while it is difficult for the human mind to picture anything beyond three dimensions, this is no problem because the computer can figure such things for us. It is to aid intuitive understanding that all the fundamental arguments concerning regression are made using the bivariate case. The student who can grasp the arguments at this level can then easily accept that the same fundamental conditions hold for models with increased numbers of variables.

2.3 Linear regression: a formal interpretation

Our general theoretical speculation regarding the relationship between an explained variable Y and an explanatory variable X is

$$Y = \alpha + \beta X$$

But in reality we cannot know this line and have to resort to a statistical model which takes into account possible unknown explanatory variables and which relates sets of observations of the variables to one another. This equation is the statistical version of the theoretical model and is designed to accommodate situations where paired observation points do not all fall upon a straight line.

$$Y_i = \alpha + \beta X_i + u_i \quad \text{where } i = 1, 2, 3, 4, \ldots, n \text{ is a} \qquad (2.2)$$
counting method of locating each set of observations

It should be obvious that this model takes each pair of observations into account.

Diagram 2.6 shows how the statistical model is designed to accommodate not only the theoretical model but also the data point deviations from that model. When $i = 1$, the statistical model becomes

$$Y_1 = \alpha + \beta X_1 + u_1$$

and this point is shown to the left of the diagram. Notice that every observation data point can be described in a like fashion by the statistical model. Also notice that we are only concerned with the vertical distance from the line.

Diagram 2.6

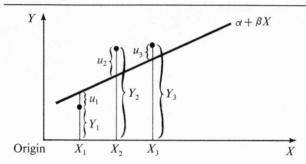

Having designed this model, we can go to the observations and some number-crunching technique to obtain estimates of the parameters α and β. This produces the estimated line

$$\hat{Y}_i = \hat{\alpha} + \hat{\beta} X_i \qquad (2.3)$$

where $\hat{Y}_i$ is the estimated value of Y_i, $\hat{\alpha}$ is the estimated value of the parameter α, and $\hat{\beta}$ is the estimated value of the parameter β. We do not need an estimate of X because it is the explanatory variable, and in the experimental conditions under which we are operating, X can be measured with precision.

Finally, we can define a *residual*, e_i, as

$$e_i = Y_i - \hat{Y}_i \qquad (2.4)$$

Thus the residual is the difference between the true Y_i and the $\hat{Y}_i$ estimate obtained from the regression calculation. Obviously e_i is an estimate of

Diagram 2.7

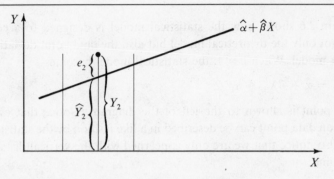

u_i. Diagram 2.7 shows the location of the estimated line within the observations from which the estimated line is calculated.

Thus we have three types of model. We have the *theoretical model* which forms the basis for our other models and delineates the relationship between the explained and explanatory variable. This model is imported into statistics by the generation of a *statistical model* whose function is to account for the theoretical thrust and link this to every observation obtained. Finally, we have to devise some method of estimating the parameters of the statistical model. All estimation techniques contain weaknesses which produce deviations from the statistical model, and an *estimator model* is produced.

2.4   Criteria for estimating the regression line

We have defined the *residual*, in any regression situation, as

$$e_i = Y_i - \hat{Y}_i$$

It would follow that if we can arrange the regression line such that Y_i is equal to $\hat{Y}_i$ then we have the most favorable line. In this situation the estimated values of the explained variable are equal to the observed values. Consequently, our aim in selecting a procedure for determining the displacement of the line should be to reduce the residual, e_i.

There are various formulations which have the property of reducing the residual component. We shall look at three of them. The first two have fatal disadvantages; the third, however, is the one we shall eventually select.

The first technique allows us to select a regression line that minimizes the sum of the residuals, that is,

$$\text{Minimize} \quad \sum (Y_i - \hat{Y}_i) \quad \text{or minimize} \quad \sum e_i$$

The difficulty with this criterion is that to satisfy it we need only to choose a line where all values of $\hat{Y}_i$ are so large that the term $\Sigma (Y_i - \hat{Y}_i)$ is a large negative number—the larger the negative number, the nearer we come to satisfying the criterion. With this flaw our selected line will be unrelated to any of the points in the data plot (see diagram 2.8). In case A the line is placed reasonably. However, if we follow this criterion we would select the line in case B. Thus the criterion has provided a line which is obviously inferior. We can disregard this criterion.

Diagram 2.8

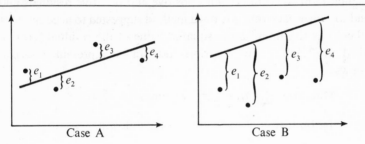

Case A   Case B

An improvement upon this first criterion is a second in which we minimize the absolute values of the residuals, i.e., we disregard the sign of the residual and just summate all of them. The line which produces the lowest total is the one that we would select. More formally we would

$$\text{Minimize} \quad \sum |(Y_i - \hat{Y}_i)| \quad \text{or equivalently, minimize} \quad \sum |e_i|$$

The difficulties produced here are different. Clearly, our regression line should take into account every observation in the sample. But using this criterion it is possible to prefer a line that does not include every observation. This is unacceptable. In omitting data points we are throwing away valuable information. The only consequence of throwing away data is the generation of incorrect estimators. Consider diagram 2.9. In scattergram A, the sum of the absolute values of the residuals equals 5. In scattergram B, the same sum equals 3. Under our second criterion we

Diagram 2.9

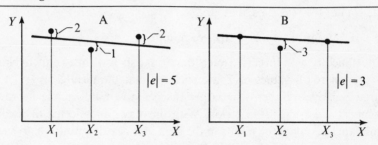

would thus select the line in B, but this line would be the same whether or not point X_3 exists. We would be unwise to select some line which does not exploit all the empirical information.

In all of this, notice that the difficulty is in dealing with the sign of the residual term. We can get both positive and negative residuals and they tend to cancel each other. A third method suggested to accommodate this difficulty is to minimize the squared value of the residual term. This is called the least-squares criterion and can be represented formally as follows:

$$\text{Minimize} \quad \sum (Y_i - \hat{Y}_i)^2 \quad \text{or minimize} \quad \sum (e_i)^2$$

This criterion turns out to be best for the following reasons:

(1) the resultant summation will be nonnegative,
(2) the resultant summation will vary directly with deviation of points from the line,
(3) the resultant summation takes every observation into account,
(4) the criterion can be theoretically justified.

2.5   The importance of the error terms in regression

It should now be very clear that the important component in regression analysis is the error term. Since the derivation of the estimator is focused upon the error term and its minimization, we should pay particular attention to the assumptions that we make about the distribution of the error terms.

The error terms will be distributed around the true regression line, as diagram 2.10 shows. By assumption all of the errors will fall somewhere

Diagram 2.10

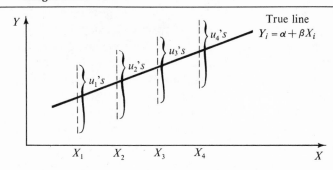

on the dashed lines. Since we are dealing with the true line, the errors are symbolized by u's. Notice initially that the width of the band within which the error terms fall is exactly the same no matter what the value of the explanatory variable.

Let us look at one particular value of the explanatory variable, say X_i. Suppose we carry out an infinite number of experiments using this value of the explanatory variable; for example, if we had spent $1000 in an infinite number of wards in our money-spent–vote-increase example. Because most of the data points would be closer to the regression line and less away from the regression line, we would expect the distribution of u_1 to be normally distributed with a mean of zero (diagram 2.11). To be complete in describing this distribution of errors around the true regression line, we can say that, as well as having a mean of zero, the distribution will also have some finite variance, which we can call σ_1^2. We can formally define the distribution of u_1 as

$$u_1 = n(0, \sigma_1^2)$$

Diagram 2.11

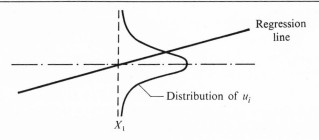

But since we have seen before that each value of the explanatory variable will generate its own set of error terms, each with its own distribution, we can provide a three-dimensional perspective to the error terms along the true regression line. Note that these will be planes equidistant from one another. The distance between them will be dependent upon the closeness of the values of the explanatory variable. One can conceive of a situation where the planes are so close as to be indistinguishable from each other. In this situation we would obtain a ridgelike structure which would have as its cross section the normal curve. This ridgelike structure would be sitting symmetrically astride the true regression line. The implication of this is that we can move away from our experimental situation because we can imagine an infinite number of potential values of the explanatory variable, any of which is as valid as the other. Thus we can take the explanatory variable as we find it instead of setting up experimental values of it. We shall discuss this later.

In diagram 2.12, each of the error distributions will have a mean of zero and a variance of σ_i^2 where i locates the value of the explanatory variable.

Diagram 2.12

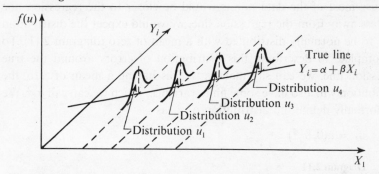

Having looked at the error term, we can now proceed to list the critical assumptions made in least-squares regression. Most of them concern the error term distribution. Thus we see, as we might have expected, it is attention to what is excluded from the regression model which occupies our effort. We could argue further that data points deviant from the line are those which will lead us to look for other variables to include in the model to reduce such error. Data that sit on the regression line are relatively uninteresting. It is deviance that promotes inquiry.

2.6   The five assumptions in regression

There are five critical assumptions in regression. We shall look at each one in turn and discuss their importance.

(1) Normality: u_i is normally distributed.

This first assumption states that the distribution of errors around the regression line is normal. Thus the error distribution is symmetrical about the regression line. In this we consider that the error term is an accumulation of a number of smaller errors in our misspecification of the model and in the gathering of the data. These small errors produce a deviation from the regression line. But these small deviations are random and additive, thereby producing the normal distribution.

It is true, however, that the characteristics of the estimators, which we discuss shortly, are not lost should the distribution of the error term be other than normal. The normality criterion is only required to yield maximum likelihood estimates of the regression parameters. We shall not be dealing with such estimates in this text but the student can consult a plethora of books on this topic.[1]

(2) Zero mean: the expected value of u_i is zero.

Assumption 2 has far more consequence for estimation accuracy than the previous assumption. Assumption 2 states that the mean of the distribution of each of the error distributions is equal to zero. That is, the mean of each of the error distributions falls on the true regression line. Thus, the errors are located symmetrically about this mean and around the regression line.

In least-squares regression the procedure attempts to locate the line on the mean of the error distribution. If the distribution is not symmetrical about the true regression line but some other value off the regression line, our estimate will reflect this. So although our slope coefficient may not be altered, the intercept coefficient will be. This bias in the intercept will be the same as the distance between the mean of the error distribution and the true regression line. When the mean of the error distribution is on the true regression line, the bias in the intercept will be zero. The two sketches in diagram 2.13 show this.

1. Kmenta, *Elements of Econometrics*, pp. 174–82, 213–15, 356–57.

Diagram 2.13

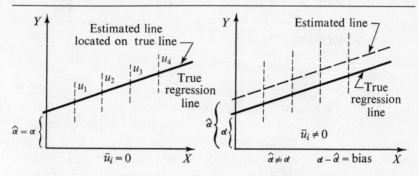

(3) Homoscedasticity: all of the variances of the error terms σ_i^2 are constant.

As diagram 2.14 shows, we require the left-hand situation rather than the right-hand one. The term "homoscedasticity" means that the variance of the error distribution for each value of the explanatory variable is constant and does not change with the value of the explanatory variable. If the variance increases as the value of the explanatory variable increases (the situation on the right), the errors at the higher values of the explanatory variable would have a larger effect upon the variance of the sampling distribution than those where the variance is small. We find that the variance of the sampling distribution of β is biased when the error distribution is heteroscedastic and consequently our inference-testing sequence will be rendered inaccurate.

Substantively, homoscedasticity means that the various factors causing

Diagram 2.14

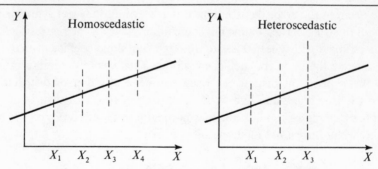

the error terms do not change over the range of values of the explanatory variable. If the variance of the error distribution does change as the explanatory variable changes, it suggests that some factor in the error component is related to the explanatory variable. The error term is not independent of the explanatory variable. As we shall see when considering the characteristics of the regression estimators, all characteristics are dependent upon an assumption that the error term and the explanatory variable are not related. In heteroscedasticity we are told that some of the unknowns in the error component are related to the known explanatory variable.

    (4) No autocorrelation of the error terms: covariance of e_i and e_j is zero $(i \neq j)$.

Verbally, assumption 4 states that no error term should be related to any preceding or successive error term. The error term at X_1 should be independent of the error term at X_2 and so on. This assumption is usually violated in time series data when the explanatory variable is observed at successive time points. It is most likely to occur as the time distance between observations decreases.

Diagram 2.15 provides some demonstration of possible difficulties. The error terms track one another as the explanatory variable value increases. But we as politometricians do not have the true regression line. All we have are the data points. The least squares regression procedure will locate a line that is distinct from the true regression line. Visually, it is hard to determine autocorrelation in the residuals of our regression output.

Diagram 2.15

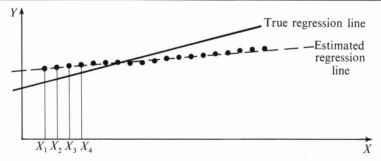

Nevertheless, there are standard techniques for detecting the presence of autocorrelation in error terms.

(5) Nonstochastic explanatory variable.

Assumption 5 is based in the experimental development of regression. The values of the explanatory variable are required to be precisely controlled. There is no allowance for any measurement error, and there is the subsidiary requirement that the variance of the explanatory variable must be finite and different from zero. This implies that we must have various different values of the explanatory variable before we can locate a line. When the variance of the explanatory variable is close to zero we have a degenerate situation as shown in diagram 2.16, at left. There are a multitude of lines which might conceivably go through these points. However, as the variance of the explanatory variable gets larger, it becomes easier to locate the regression line.

Diagram 2.16

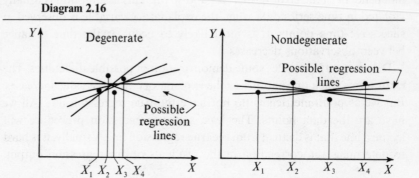

These are the five assumptions of classical regression. As we shall see as we progress through the book, reality almost always violates one or all of these assumptions. This being so, two problems confront us. First, we need methods to detect when the assumptions have been violated. Obviously our best estimate of the error term u_i is our residual e_i. Analysis both visual and statistical must take place on all residuals—before the regression coefficients are even glanced at!

Second, if the assumptions have been violated, we need techniques for overcoming the various inaccuracies that such violations may have brought into our estimation procedures.

2.7 The normal equations

We can use the least-squares criterion to explore possible mathematical maneuvers to obtain the actual estimators. It is the hallmark of all mathematical procedures to be concerned with the underlying assumptions before developing the operative function. This is a pattern of behavior worth emulating. Our aim is to determine the estimator, but this will be the last thing that we look at in any real situation. We shall initially be concerned with those factors which determine whether we have violated any of the regression assumptions. Only after having satisfied ourselves on that score will we allow ourselves a peek at the actual value of the estimator.

Our least-squares criterion prescribes a wish to minimize the sum of the squared residuals, that is,

$$\text{Minimize } \sum e_i^2 \tag{2.5}$$

But we know that $e_i = Y_i - \hat{Y}_i$ and that $\hat{Y}_i = \hat{\alpha} + \hat{\beta}X_i$. Thus our criterion implies that we should

$$\text{Minimize } \sum (Y_i - \hat{\alpha} - \hat{\beta}X_i)^2$$

If we square this term we obtain

$$\text{Minimize } \sum (Y_i^2 - 2Y_i\hat{\alpha} - 2Y_iX_i\hat{\beta} + \hat{\alpha}^2 + 2\hat{\alpha}\hat{\beta}X_i + \hat{\beta}^2X_i^2)$$

In using any minimizing procedure, it is necessary to partially differentiate the equation to be minimized with respect to the parameters of interest. We have two such parameters, $\hat{\alpha}$ and $\hat{\beta}$, and so we must differentiate with respect to both of them. On completing this for the above equation for the two parameters, we obtain two equations. These are called the *normal equations* and are presented below.

$$\sum Y_i = n\hat{\alpha} + \hat{\beta}\sum X_i \qquad \text{Normal equation I} \quad (2.6a)$$

$$\sum Y_iX_i = \hat{\alpha}\sum X_i + \hat{\beta}\sum X_i^2 \qquad \text{Normal equation II} \quad (2.6b)$$

In these equations, the only unknowns are $\hat{\alpha}$ and $\hat{\beta}$. All the other terms are obtainable from the data. For instance $\sum Y_i$ is all of the Y observations added together: n is the sample size. And $\sum X_iY_i$ is the sum of each individual X_i multiplied by the corresponding Y_i. When all of these values are inserted from the available data we have two simultaneous equations, both of which contain two unknowns, $\hat{\alpha}$ and $\hat{\beta}$. We can solve the equations for these unknowns, thereby obtaining the parameters of our line derived in accordance with our criterion of least squares.

Let us solve these two equations in their general form and show that the resulting equations are far more agreeable to deal with. We obtain the following familiar relationships:

$$\hat{\beta} = \frac{\sum\limits_{i=1}^{n} X_i Y_i - \dfrac{\sum\limits_{i=1}^{n} X_i \sum\limits_{i=1}^{n} Y_i}{n}}{\sum\limits_{i=1}^{n} X_i^2 - \dfrac{\left(\sum\limits_{i=1}^{n} X_i\right)^2}{n}} \qquad (2.7a)$$

$$\hat{\alpha} = \frac{\sum\limits_{i=1}^{n} Y_i}{n} - \frac{\hat{\beta} \sum\limits_{i=1}^{n} X_i}{n} \qquad (2.7b)$$

We can now try out our calculation equations using the data from our example of money spent–percentage increase in vote, for which data are presented in table 2.1.

Table 2.1

	The data		Provided for the calculations		
i	Y_i	X_i	Y_i^2	X_i^2	$X_i Y_i$
1	2.8	1,000	7.84	1,000,000	2,800
2	3.0	1,000	9.00	1,000,000	3,000
3	3.1	1,000	9.61	1,000,000	3,100
4	5.7	2,000	32.49	4,000,000	11,400
5	6.0	2,000	36.00	4,000,000	12,000
6	6.2	2,000	38.44	4,000,000	12,400
7	8.5	3,000	72.25	9,000,000	25,500
8	9.0	3,000	81.00	9,000,000	27,000
9	9.1	3,000	82.81	9,000,000	27,300
10	11.6	4,000	134.56	16,000,000	46,400
11	11.9	4,000	146.61	16,000,000	47,600
12	11.9	4,000	146.61	16,000,000	47,600
13	14.7	5,000	216.09	25,000,000	73,500
14	15.2	5,000	231.04	25,000,000	76,000
15	15.1	5,000	228.01	25,000,000	75,500
16	17.6	6,000	309.76	36,000,000	105,600
17	18.0	6,000	324.00	36,000,000	108,000
18	18.2	6,000	331.24	36,000,000	109,200
Total	187.6	63,000	2,427.36	273,000,000	813,900

When we insert the summations at the bottom of each column into equations (2.7a) and (2.7b) we obtain the estimates of $\hat{\alpha}$ and $\hat{\beta}$.

$$\hat{\beta} = \frac{813,900 - \dfrac{187.6 \times 63,000}{18}}{273,000,000 - \dfrac{(63,000)^2}{18}} = \frac{2.9962}{1,000}$$

$$\hat{\alpha} = \frac{187.6}{18} - \frac{2.9962}{1,000} \times \frac{63,000}{18} = -0.064$$

Since we derived our calculating equations from the least-squares criteria, these estimates are the least-square estimates of the slope and intercept of the regression line. Thus our estimating equation for the money-spent–increase-in-vote situation is

$$Y_i = -0.064 + \frac{2.9962}{1,000} X_i$$

Notice how our eyeball guestimate of the parameters was not too unreasonable. Not only is the least squares criterion mathematically rewarding, but it also seems to agree with what we would have visually predicted.

2.8   Another intuitive discussion of the regression slope

Since the regression slope is directly linked to the substantive implications of the relationship between the explained and the explanatory variables, let us have another look at the slope coefficient using its calculating equation. The equation for the regression slope estimator is

$$\hat{\beta} = \frac{\sum X_i Y_i - \dfrac{\sum X_i \sum Y_i}{n}}{\sum X_i^2 - \dfrac{(\sum X_i)^2}{n}}$$

This can be rearranged to

$$\hat{\beta} = \frac{\sum (X_i - \bar{X})(Y_i - \bar{Y})}{\sum (X_i - \bar{X})^2} \cdot \frac{n}{n} \tag{2.8}$$

The equation for $\hat{\alpha}$,

$$\hat{\alpha} = \frac{\sum Y_i}{n} - \hat{\beta}\frac{\sum X_i}{n}$$

can be rearranged to

$$\hat{\alpha} = \bar{Y} - \hat{\beta}\bar{X} \quad \text{where } \bar{Y} \text{ is the mean of the } Y \text{ distribution} \qquad (2.9)$$
$$\bar{X} \text{ is the mean of the } X \text{ distribution}$$

This equation tells us that the regression line goes through the intersection of the means of the two distributions. Let us divide up the scattergram space horizontally and vertically through the intersection of the two means (diagram 2.17).

Diagram 2.17

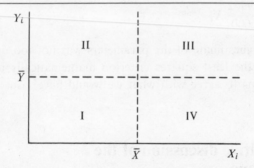

Notice that all the observations to the left of the vertical line through the intersection of the means will produce values of $(X_i - \bar{X})$ which are negative. All those to the right of the vertical line will produce values of $(X_i - \bar{X})$ which are positive. Similarly, observations that fall above the horizontal line will produce values of $(Y_i - \bar{Y})$ which are positive, and those which fall below the horizontal line will produce values of $(Y_i - \bar{Y})$ which are negative. When we multiply these various positives and negatives together as we do in the numerator of the revised equation for $\hat{\beta}$, we get a positive product in quadrants I and III and a negative product in quadrants II and IV. Since the denominator in the equation is always positive, we can see that the sign of the slope coefficient is entirely dependent upon the sign of the numerator. The regression slope will be positive when the observations fall in quadrants I and III and negative when the observations fall in quadrants II and IV. Knowing that the regression line must go through the intersection of the means of the two distributions, we see that when the observations are in the I and III quadrants the line has to go from the bottom left to the top right of the scattergram space, and when the observations are all in the quadrants II and IV, the slope must go from the top left to the bottom right of the scattergram space. Thus the intuitive interpretation we obtained from the equations derived from our least squares regression criterion jibes with our original intuitive discussion of the regression slope—further comfort to those selecting this criterion for locating a regression line.

Finally, in this intuitive recapitulation, the equation

$$\hat{\beta} = \frac{\sum (X_i - \bar{X})(Y_i - \bar{Y})}{\sum (X_i - \bar{X})^2} \cdot \frac{n}{n}$$

can be represented verbally by the relationship

$$\frac{\text{covariance of } X \text{ and } Y}{\text{variance of } X} \tag{2.10}$$

This is because the covariance of X and Y is defined by

$$\frac{\sum (X_i - \bar{X})(Y_i - \bar{Y})}{n} \tag{2.11}$$

and the variance of X is defined by

$$\frac{\sum (X_i - \bar{X})(X_i - \bar{X})}{n} \quad \text{or} \quad \frac{\sum (X_i - \bar{X})^2}{n}$$

If X is related to Y, then they will covary and the quantity in the numerator will be greater than zero. Thus, depending upon the variance of X, $\hat{\beta}$ will be some nonzero value, which we would interpret as measuring some relationship between the two variables. However, if X and Y do not covary the numerator will be zero and $\hat{\beta}$ will also be zero, whatever the variance of X. In this case we would obtain a value of $\hat{\beta} = 0$ and conclude that there is no relationship between the two variables. Again the estimator of the regression slope derived from the least-squares criterion has conformed with our intuitive notion about the relationship between explained and explanatory variables, given zero and nonzero values for $\hat{\beta}$.

2.9   The characteristics of the least-square estimators

In Chapter 1 we discussed various desirable characteristics of any estimator. These included unbiasedness, efficiency, and consistency. The least-squares regression coefficient has all these properties, provided that the residual, which as we know represents the unknown explanatory variables left out of the model, is uncorrelated with the included explanatory variable X_i. Thus, provided that e_i is distributed independently of X_i the estimator is unbiased, and for any given sample size, the mean value of the sampling distribution of the estimator is efficient and the variation of the sampling distribution around the true value of β is less than any other competing estimator. And as the sample size of the sampling distribution is increased, all of the values of $\hat{\beta}$ move closer to the true value of β. This is the property of consistency. $\hat{\beta}$ is also a linear function of the estimators. In short, $\hat{\beta}$ is BLUE. The estimator α also has these properties.

As well as these characteristics, the least-squares regression estimators also satisfy the more easily met asymptotic or large-sample properties. For those readers interested in the mathematical arguments for the above assertions, there are a number of books available.[2]

2.10   The goodness of fit of the regression line—correlation coefficient

Despite its prolific use in political science research, the Pearson product-moment correlation coefficient is of less use to the theoretician than the regression coefficient. Because it requires no *a priori* thinking concerning direction of influence in any relationship, the correlation coefficient is an indolent measure of the magnitude of the relationship between two variables. As Malinvaud has noted,

> Around 1930, when the systematic study of dependence between economic quantities was first undertaken, it was based on the use of correlation coefficients. Today most studies in econometrics are based on regression analysis and correlation coefficients are virtually ignored.[3]

2. Kmenta, *Elements of Econometrics*, pp. 162–64, and asymptotic variance, p. 182.

3. E. Malinvaud, *Statistical Methods of Econometrics* (Chicago, Ill.: Rand-McNally, 1966), p. 25.

It is also indicative of the waning importance of the correlation coefficient that one of the standard textbooks on econometrics, by Goldberger,[4] does not even index the term correlation. Presumably, as econometrics has developed so will politometrics, and we shall eventually see an eclipse of the importance of the correlation coefficient as a measure of relationship.

Why then is the correlation coefficient so fashionable in political science? Because with this coefficient the researcher need not trouble deciding whether X causes Y or Y causes X; whether they both cause each other simultaneously; or indeed, whether they are theoretically linked at all since they may both be caused by a prior variable. Finally, the correlation coefficient does not require a nonstochastic variable; thus it is more suited to the nonexperimental character of political phenomenon. The requirements for correlation are not as rigorous as those for regression, with the result that this measure has been used extensively by other writers.

There are two kinds of research into relationships between variables. The first is a *scalar analysis* of relationships and is concerned exclusively with determining the *magnitude* of the relationship. The second is called a *vector analysis* of relationships and deals with both *the magnitude and the direction* of a relationship between variables. While correlation is concerned with scalar analysis, regression (with its *a priori* determination of the explained and explanatory variables) is concerned with vector relationships. In terms of theoretical advancement, vector analysis must be preferred to scalar analysis.

The correlation coefficient, nevertheless, is linked mathematically with the regression coefficient. When the regression coefficient is zero, so too is the correlation coefficient, and the correlation coefficient will have the same sign. Since both are so closely related, is there any way in which we can exploit the correlation coefficient? Yes. The regression coefficients tell us *how* two variables are related. They determine the relationship between the cause and the effect. The correlation coefficient can tell us *how well* the variables are related. It is in this subsidiary role that the correlation coefficient can best be exploited. So while regression will contain theoretical information concerning the magnitude and direction of influence between variables, the correlation coefficient (and other

4. A. S. Goldberger, *Econometric Theory*.

correlational devices we shall later develop) will determine the "goodness of fit" of our regression model to the data. Thus, regression is concerned with the development of laws relating variables, and correlation is a qualificatory device telling whether the obtained law is good, bad, or indifferent.

2.11   Correlation: a preliminary intuitive discussion

Since we have given the correlation coefficient this subsidiary role in our maneuvers, we will assume that we have already developed some estimate of the regression line. Suppose that the regression line is that shown in part A of diagram 2.18. This line is our law relating the explained variable Y to the explanatory variable X. If this law is perfect, all of the data points we get from paired observations of the two variables will fall exactly on the regression line. Simply, all of the variance of Y is explained by X. In this situation, the absolute value of the correlation coefficient will be one.

Diagram 2.18

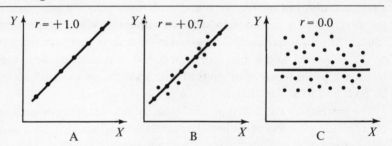

Suppose, however, that our law relating X and Y is not perfect but leaves a significant amount of the variation of Y unexplained by X. We would expect our data points to move away from the regression line. The vertical distance of the data points from the regression line, as we have seen before, is due to factors such as misspecification of the model or measurement error. In this situation the correlation will be somewhat less than unity, and in some intuitive manner the difference between unity and this lower value of the correlation coefficient is a measure of the inadequacy of our model. This situation is shown in part B.

Let us finally suppose that there is no relationship between X and Y. In this situation we would obtain a slope estimate of zero, signaling us that the regression line is horizontal. In this situation our law is completely wrong (but still interesting!) and the correlation coefficient will reflect this by attaining a value close to zero. This is shown in part C.

Symmetrically, if there is an inverse relationship between Y and X, such that as X increases Y decreases, and the law is perfect between the variables, then the correlation coefficient will reflect this by attaining a value of -1. As the law becomes less perfect the absolute value of the coefficient will decrease but still retain its negative character. The coefficient will diminish in its absolute value until it finally arrives at zero. In this situation, as before, there will be no relationship between the explained and explanatory variable. Diagram 2.19 shows these situations.

Diagram 2.19

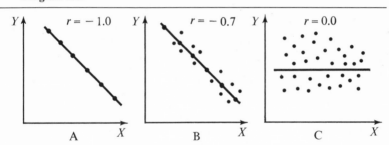

So, in summary, the absolute value of the correlation coefficient is an indication of the goodness of the regression law. At an absolute value of unity, the regression law is perfect and the explanatory variable accounts for all the variance in the explained variable. As the absolute value of the correlation falls below unity, this indicates the limitations of our regression law. We will finally reach a point where the absolute value of the correlation coefficient is zero. At this point our regression law is hopeless (but again, still theoretically interesting).

2.12   Calculating the correlation coefficient

Since we are less interested in the correlation coefficient, we shall simply state the equation.

$$r = \frac{\sum (Y_i - \bar{Y})(X_i - \bar{X})}{\sqrt{\sum (Y_i - \bar{Y})^2 \sum (X_i - \bar{X})^2}}$$

(2.12)

Verbally this equation can be represented by

$$r = \frac{\text{covariance of } X \text{ and } Y}{(\text{standard deviation } X)(\text{standard deviation } Y)} \tag{2.13}$$

Notice the similarity to the verbal version of the regression equation for the slope

$$\beta = \frac{\text{covariance of } X \text{ and } Y}{\text{variance of } X} \tag{2.14}$$

The correlation coefficient equation can be simplified for calculation purposes to the following

$$r = \frac{\dfrac{\sum X_i Y_i}{n} - \left(\dfrac{\sum X_i}{n}\right)\left(\dfrac{\sum Y_i}{n}\right)}{\sqrt{\dfrac{\sum Y_i^2}{n} - \left(\dfrac{\sum Y_i}{n}\right)^2}\sqrt{\dfrac{\sum X_i^2}{n} - \left(\dfrac{\sum X_i}{n}\right)^2}} \tag{2.15}$$

We can obtain all of the components of the right-hand side of the equation from the data. We shall use the money spent–percentage increase in vote example. The reader can check that $\sum X_i = 63,000, \sum Y_i = 187.6, \sum X_i Y_i = 813,900, \sum X_i^2 = 273,000,000$, and $\sum Y_i^2 = 2427.36$. Putting these into the calculation equation, we get

$$r = \frac{\dfrac{813,900}{18} - \left(\dfrac{63,000}{18}\right)\left(\dfrac{187.6}{18}\right)}{\sqrt{\dfrac{2427.36}{18} - \left(\dfrac{187.6}{18}\right)^2}\sqrt{\dfrac{273,000,000}{18} - \left(\dfrac{63,000}{18}\right)^2}} = 0.9991$$

The reader should notice that the correlation coefficient is positive, as we would would have predicted from knowing that the regression slope was positive. So far, little has been added to our knowledge. The absolute value of the correlation coefficient is 0.9991. This means that although our regression model of the situation is not perfect, it almost is. This is information which the correlation coefficient has added. And it is obviously useful information.

The reader will also have seen that the correlation coefficient, unlike the regression coefficient, has limits. These are $+1$ and -1. While the value of the regression-slope coefficient is dependent upon the units with which we measure the variables, the correlation coefficient is not. The unitless characteristic is produced by the division of the covariance of the variables by the product of their standard deviations. This facilitates comparison between various laws which may very well be substantively different. Thus the goodness or badness of the law is unrelated to its substantive content and related only to the way in which we have modeled the real situation.

2.13 Relationship between the regression slope and the correlation coefficient

Comparison of the equations for the regression slope and the correlation coefficient allows us to develop a mathematical relationship between the two. The equation for the correlation coefficient is

$$r = \frac{\sum (Y_i - \bar{Y})(X_i - \bar{X})}{\sqrt{\sum (Y_i - \bar{Y})^2} \sqrt{\sum (X_i - \bar{X})^2}} \cdot \frac{n}{n}$$

If we multiply top and bottom of this equation by the standard deviation of X, we obtain the equation

$$r = \frac{\sum (Y_i - \bar{Y})(X_i - \bar{X})}{\sum (X_i - \bar{X})^2} \cdot \frac{\sqrt{\sum (X_i - \bar{X})^2}}{\sqrt{\sum (Y_i - \bar{Y})^2}} \cdot \frac{n}{n}$$

but we know that

$$\hat{\beta} = \frac{\sum (Y_i - \bar{Y})(X_i - \bar{X})}{\sum (X_i - \bar{X})^2} \cdot \frac{n}{n}$$

and so, substituting this, we obtain an equation with both r and $\hat{\beta}$ in it, thus demonstrating the following mathematical relationship:

$$r = \hat{\beta} \cdot \frac{SD(X)}{SD(Y)} \tag{2.16a}$$

or

$$\hat{\beta} = r \cdot \frac{SD(Y)}{SD(X)} \tag{2.16b}$$

We see that the relationship between the two coefficients is provided by a ratio of the standard deviation of X to the standard deviation of Y.

Mathematically, it will be obvious that since both the standard deviations of X and Y must be positive, the sign of r will be the same as the same as the sign of $\hat{\beta}$. Also, if $\hat{\beta}$ is zero, r will equal zero. This was stated in the earlier intuitive discussion of r. But for completeness we can also show this visually. Let us take the original equation for the correlation coefficient

$$r = \frac{\sum (Y_i - \bar{Y})(X_i - \bar{X})}{\sqrt{\sum (Y_i - \bar{Y})^2} \sqrt{\sum (Y_i - \bar{Y})^2}} \cdot \frac{n}{n}$$

Looking at the denominator, we see that this will always be positive, so that the sign of r will always be dependent upon the numerator. As before, we can divide the scattergram space into four quadrants by drawing horizontal and vertical lines to pass through the mean of Y and the mean of X, respectively (diagram 2.20). Any observations that occur to the right of the vertical line will render the value of the term $\sum (X_i - \bar{X})$ positive, and any observations to the left of the line will provide negative values for the term. Similarly, any observations above the horizontal line will render the term $\sum (Y_i - \bar{Y})$ positive, while those below the line will produce negative values. Thus, if we consider the scattergram space quadrant by quadrant, we see that in quadrants I and III the numerator in our correlation coefficient equation will be positive, making our correlation coefficient positive. And if observations occur predominantly in quadrants II and IV, our numerator and consequently our correlation coefficient would be negative. All of this is in line with similar arguments concerning the location of observations we made when considering the regression slope.

Diagram 2.20

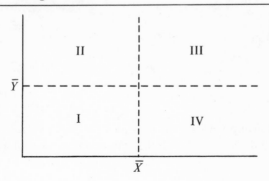

Further readings

Substantive

There is a plethora of studies in political science journals which use bivariate regression and correlation. The student can easily find these. Some special readings on this include the following.

Gurr, T. R. 1972. *Politimetrics*. Englewood Cliffs, N.J.: Prentice-Hall.
Kirkpatrick, S. A. 1974. *Quantitative Analysis of Political Data*. Columbus, Ohio: Merrill.
Tufte, E. R. 1970. *Quantitative Analysis of Social Problems*. Reading, Mass.: Addison-Wesley.
Tufte, E. R. 1974. *Data Analysis for Politics and Policy*. Englewood Cliffs, N.J.: Prentice-Hall.

Statistical

Johnston: pp. 8–35.
Kmenta: pp. 201–13.
Wonnacott and Wonnacott: pp. 1–19, 103–16.

3

various testing sequences

Developing the parameters and measures of goodness of fit of the line may only be an initial step in a whole process. Following this could be a need for testing hypotheses about the parameters or developing confidence intervals around them. Eventually, one might very well be concerned with predicting values of the explained variable from values of the explanatory variable. All of these are important maneuvers. This chapter will be divided into three parts to discuss such matters. The first part will deal with hypothesis testing, the second with developing confidence intervals, and the third with predictions about the explained variable.

3.1 Hypothesis testing

An important part of any potential theoretical advancement is the testing of hypotheses. Such testing normally follows the classical testing sequence:

(1) choosing the significance level,
(2) determination of the acceptance/rejection level,
(3) selection and generation of the statistic,
(4) the decision.

Our major concern here will be with selection and generation of the test statistic. But first, a recapitulation of the standard form of the test statistic. The distribution of the standardized normal variable is

$$Z = \frac{\text{estimated parameter} - \text{mean of the sampling distribution}}{\sqrt{\text{variance of the sampling distribution}}}$$

and Z will be distributed with a mean of zero and a variance of unity. We can exploit the properties of the standardized normal curve in our testing sequence. Thus in the case of the two parameters we estimated, the test statistics will be

$$Z = \frac{\hat{\alpha} - \alpha}{\sqrt{\sigma_{\hat{\alpha}}^2}} \tag{3.1}$$

$$Z = \frac{\hat{\beta} - \beta}{\sqrt{\sigma_{\hat{\beta}}^2}} \tag{3.2}$$

where $\sigma_{\hat{\alpha}}^2$ and $\sigma_{\hat{\beta}}^2$ are the variances of the respective sampling distributions.

In both of these equations the only unknowns are the variances of the sampling distribution of α and β. These can be calculated from the relationships that follow. The reader can go to a selection of books to uncover how the relationships are derived.[1]

$$\sigma_{\hat{\alpha}}^2 = \frac{\sigma^2}{n} \cdot \frac{\sum X_i^2}{\sum (X_i - \bar{X})^2} \tag{3.3}$$

$$\sigma_{\hat{\beta}}^2 = \frac{\sigma^2}{\sum (X_i - \bar{X})^2} \tag{3.4}$$

Thus equations (3.1) and (3.2) become respectively

$$Z = \frac{\hat{\alpha} - \alpha}{\sqrt{\dfrac{\sigma^2}{n} \cdot \dfrac{\sum X_i^2}{\sum (X_i - \bar{X})^2}}} \tag{3.5}$$

and

$$Z = \frac{\hat{\beta} - \beta}{\sqrt{\dfrac{\sigma^2}{\sum (X_i - \bar{X})^2}}} \tag{3.6}$$

1. R. J. Wonnacott and T. H. Wonnacott, *Econometrics*, pp. 19–21.

Notice that in both equations (3.3) and (3.4), the larger the value of the component $\Sigma (X_i - \bar{X})^2$ the smaller will be the variance of the estimator. This increases the efficiency of our estimator—a desirable trait. To increase this component we need values of the explanatory variable highly dispersed around the mean of the distribution of the explanatory variable. We see that in the best possible situation we should have a wide range of values for the explanatory variable, because when all of the values of the explanatory variable are the same, the component $\Sigma (X_i - \bar{X})^2$ will be zero and the variance of our estimators infinite and inefficient. This is called a degenerate situation. The characteristics of our estimator are enhanced by a wide dispersion in the values of the explanatory variable (diagram 3.1). The left-hand diagram depicts almost degenerate experimental situation.

Diagram 3.1

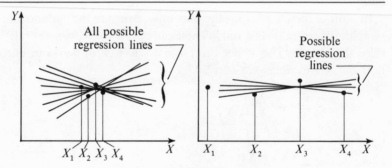

The second noticeable feature of variances of the estimators is that they are both dependent upon the variance of the error terms, σ^2. As the variance of the error terms increases, so will the variance of the sampling distribution of the estimators. Again a diagram will aid in understanding this. In the left-hand part of diagram 3.2, the variance of the error terms is large; thus the infinity of different lines that we can locate through these data points will have a wide range. In the right-hand part, the variance of the errors is small and the infinity of possible lines will have a very small range.

The variance of the error terms is normally something over which we have little control. It reflects, in some way, the adequacy of our model. But the dispersion of our explanatory variable values we may sometimes

Diagram 3.2

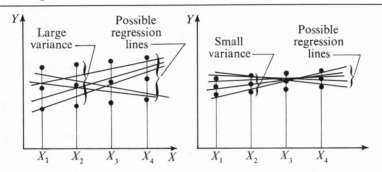

be able to manipulate (unfortunately, very rarely in political science!). These equations can give us guidance in setting up experiments.

In both equations the only component we have not discussed is σ^2. This is the variance of the error term distribution. In knowing that the variance of the error term is constant whatever the value of the explanatory variable, we can collapse all error distributions for various values of the explanatory variable into one distribution (diagram 3.3).

Diagram 3.3

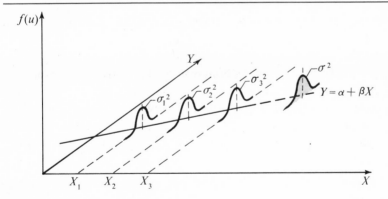

Unfortunately, in most situations, σ^2 is unknown and has to be estimated using the residuals from our parameter estimation. It was stated earlier that the best estimator of u_i is e_i. We can get e_i from the equation $e_i = Y_i - \hat{Y}_i$, but in estimating e_i we have already made use of two estimators, $\hat{\alpha}$ and $\hat{\beta}$. We therefore lose two degrees of freedom. Then σ^2

is estimated from the equation

$$\sigma^2 = \frac{\sum e_i^2}{n - K} \tag{3.7}$$

where K is the number of variables in the regression. In our bivariate situations, $K = 2$.

In using this estimator for the variance of the error terms, we have introduced bias into our procedure, particularly with small samples. To get over this difficulty, instead of using the Z-distribution we exploit the t-distribution with $n - K$ degrees of freedom. So equations (3.5) and (3.6) become, for the intercept,

$$t = \frac{\hat{\alpha} - \alpha}{\sqrt{\dfrac{\sum e_i^2}{n(n - K)}\left(\dfrac{\sum X_i^2}{\sum (X_i - \bar{X})^2}\right)}} \quad \text{with } n - K \text{ degrees} \atop \text{of freedom} \tag{3.8}$$

and, for the slope,

$$t = \frac{\hat{\beta} - \beta}{\sqrt{\dfrac{\sum e_i^2}{(n - K) \sum (X_i - \bar{X})^2}}} \quad \text{with } n - K \text{ degrees} \atop \text{of freedom} \tag{3.9}$$

Having considered the two estimators in tandem thus far, let us now take each of them individually.

3.1.1 Testing hypotheses concerning the regression slope

In theory building, hypotheses concerning the slope coefficients are usually the most important. There are basically two types of hypothesis to test—those that maintain that the slope coefficient is zero and those that maintain that it is equal to some nonzero constant. In the first situation the hypothesis is

$$H_0 : \beta = 0$$
$$H_A : \beta \neq 0$$

In the null-hypothesis we are asserting that the true slope coefficient is zero and thus no theoretical relationship exists between Y and X. If this null-hypothesis is rejected then we accept the alternative hypothesis that $\beta \neq 0$ and that there is a relationship between the explained and explanatory variable.

Let us use our example to test such a hypothesis. Our null-hypothesis states that there is no relationship between the amount of money spent in any ward and an increase in percentage of the vote. The alternative hypothesis argues for a relationship between money spent and vote increase. In deriving our test statistic (equation 3.9) we require $\Sigma\, e_i{}^2$ and $\Sigma\,(X_i - \bar{X})^2$. We saw earlier that $\Sigma\, e_i{}^2 = 0.8513$, and we can compute $\Sigma\,(X_i - \bar{X})^2 = 52{,}500{,}000$. Putting all of this into the appropriate equation, we obtain

$$t = \frac{0.002996 - 0.0}{\sqrt{\dfrac{0.8513}{(18 - 2)(52{,}500{,}000)}}} = 94.125 \quad \text{with } 18 - 2 = 16 \text{ degrees of freedom}$$

Depending upon the critical t value we select, we either accept or reject the null-hypothesis. If we use critical $t = 1.746$, which is the one-tailed 0.05 significance level, our test statistic is such that we reject the null-hypothesis and accept the alternative. We have therefore inferred from our sample that there is a relationship between money spent and vote increase.

Let us now test a hypothesis concerning a specific value of β other than zero. Suppose we have established for other wards that the regression of money spent on percentage increase in vote is equal to 0.005, and we are curious to see whether the regression slope in our 18 wards is significantly different from this value of 0.005. The following hypotheses would be developed:

$$H_0: \beta = 0.005, \quad H_A: \beta \neq 0.005$$

To test this hypothesis we set up the test statistic

$$t = \frac{0.002996 - 0.005}{\sqrt{\dfrac{0.8513}{(18 - 2)(52{,}500{,}000)}}} = -62.96 \quad \text{with 16 degrees of freedom}$$

And with a critical t value of 1.746 and 16 degrees of freedom, we again reject the null-hypothesis and accept the alternative. We have inferred that our regression for the 18 wards could not have come from the same population as that obtained from the other wards.

As we have seen in the previous chapter, there is a mathematical relationship between regression and correlation coefficients. In most cases in political research this relationship is used in testing hypotheses about relationships between variables. We have seen that when $\hat{\beta} = 0$, the correlation coefficient r is equal to zero. In testing whether a particular sample correlation could possibly have come from a population with a correlation of zero, we are at the same time testing whether there is any relationship between the two variables in the correlation. This is exactly the same as testing whether $\hat{\beta} = 0$. In this situation the hypothesis we test is

$$H_0: r = \rho = 0$$
$$H_A: r \neq \rho = 0$$

where ρ is the population correlation parameter and largely unknown. The test statistic for this is the common

$$t = \frac{r}{\sqrt{\dfrac{1 - r^2}{n - 2}}} \quad \text{with } n - 2 \text{ degrees of freedom} \tag{3.10}$$

When we use the data from our sample to test a hypothesis of this kind we develop the equation

$$t = \frac{0.991}{\sqrt{\dfrac{1 - 0.991^2}{18 - 2}}} = 94.125$$

Notice that this t value is exactly the same as that obtained when we tested a similar hypothesis using the slope parameter. We would expect this to be the case.

However, there are two important considerations in the use of the correlation coefficient for inference testing, particularly when we test differences between coefficients. The first is a statistical difficulty, while the second has more substantive import.

The correlation coefficient is designed such that its maximum and minimum values are $+1$ and -1. In testing any statistic, our assumptions are that the underlying sampling distribution is normal or that it approximates normality. But normal curves go to $+\infty$ and $-\infty$, not $+1$ and -1. Thus when we test correlation coefficients we map a truncated distribution onto one that goes to infinity in both directions.

When the correlation coefficients are around the zero point the in-accuracies that such mapping generates are minimal, as the left part of diagram 3.4 shows. However, when the correlation coefficients are some way from zero, we obtain the skewed curve shown at right in diagram 3.4.

Diagram 3.4

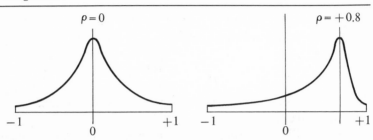

To surmount this problem we can use the Fisher transformation.[2] Fisher determined that his transformation, which is given by the relation-ship

$$z_\rho = 1.151 \log_{10} \frac{1 + r}{1 - r} \tag{3.11}$$

where r is a Pearson product-moment correlation, is normally distributed with a mean of z_ρ and a standard deviation of $1/\sqrt{(n - 3)}$. Thus our test statistic becomes

$$Z = \frac{z_r - z_\rho}{\sqrt{\dfrac{1}{n - 3}}} \tag{3.12}$$

where z_r is the Fisher transformation of our sample correlation and z_ρ is the Fisher transformation of our hypothesized population parameter. Let us take the following example by way of clarifying the use of this procedure. Suppose from a sample of 19 observations we obtain a correlation of 0.5. And suppose further that we hypothesize that this sample is from a population having a correlation of 0.8. Our hypothesis is

$$H_0 : r = \rho = 0.8$$
$$H_A : r \neq \rho = 0.8$$

2. A. E. Waugh, *Statistical Tables and Problems* (New York: McGraw-Hill, 1952), pp. 40–41.

Since the transformation is normal, we can select a Z normal critical value, and at the 0.05 level this will be 1.65. Our test statistic will be

$$Z = \frac{0.5493 - 1.0986}{\sqrt{\dfrac{1}{19 - 3}}} = -2.1972$$

The values in the numerator come from the Fisher transformation of 0.5 and 0.8. These are presented below.

$$z_r = 1.151 \log_{10} \frac{1 + 0.5}{1 - 0.5} \quad z_\rho = \log_{10} \frac{1 + 0.8}{1 - 0.8}$$

Thus at this significance level we reject the null-hypothesis and accept the alternative.

Using the Fisher transformation gets us out of the difficulty of mapping a truncated correlation coefficient onto a normal curve.

A second consideration when testing between two correlation coefficients is that we are not testing whether there is a substantive difference between applicability of the same law in two different empirical situations, but whether the goodness of the law in one situation is better or worse than in the other situation.

Let me use an example to make this more clear. One of the laws of motion is $v = u + gt$, where v is the velocity of an object under gravity, u is the initial velocity of the object, t is the time elapsed since the object was released and g is a parameter reflecting gravity. Now suppose that we carry out a series of experiments on both the earth and the moon. On both bodies we drop some object from various heights and compute the time between release and impact with the ground and the object's velocity on impact. In both cases we regress t on v and in the case of the earth obtain an estimate for g of about 32.2 while on the moon our estimate is considerably less, at around 5. Obviously the disparity between these two estimates of the value of g is very significant both statistically and substantively. We would infer something about the relative sizes of the earth and the moon affecting this gravitational parameter. Suppose, however, we test the difference between the correlation coefficients that we obtain from our experiments. Obviously, since the laws of motion are extremely good laws, both correlation coefficients would be high. We would really be testing if there was any significant difference between two high correlation coefficients. We would probably find that there was not.

But all we can conclude from this is that the laws of motion are equally good whether we are on earth or on the moon. In no way, using this test, are we allowed a glimpse at the obvious substantive differences between the parameters of the law as it operates on each body.

3.1.2 Testing hypotheses about the intercept coefficient

Theoretically, the most interesting aspect of the intercept coefficient is whether it is different from zero. If it is zero, the regression line goes through the origin of our scattergram; when X is zero, Y is zero. This may be substantively interesting.

To test if the intercept is zero we exploit the test statistic developed for the intercept coefficient (equation 3.9)

$$t = \frac{\hat{\alpha} - \alpha}{\sqrt{\dfrac{\sum e_i^2}{n(n-2)} \cdot \dfrac{\sum X_i^2}{\sum (X_i - \bar{X})^2}}} \quad \text{with } n - 2 \text{ degrees of freedom}$$

Using our money spent–vote increase example and the following hypotheses

$$H_0 : \alpha = 0$$
$$H_A : \alpha \neq 0$$

we would obtain the test statistic

$$t = \frac{-0.064 - 0.0}{\sqrt{\dfrac{0.8513}{(18)(16)} \cdot \dfrac{273,000,000}{52,500,000}}} = -0.516 \quad \text{with 16 degrees of freedom}$$

Using again the critical level of $t_c = 1.746$, we would accept the null-hypothesis and reject the alternative hypothesis. Thus we infer that the regression line in our example goes through the origin, suggesting that if you don't spend money in any wards there will be no increase in percentage of the vote.

There are some situations where a researcher may want to determine whether an intercept coefficient is different from some particular value other than zero. Diagram 3.5 sketches such a situation. In this situation we have regressed time on government spending. However, at time T it would appear that there is a step increase in the overall amount of government spending. Thus we might want to determine whether point P is

Diagram 3.5

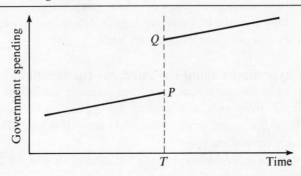

statistically significantly different from point Q. If we regress time on government spending, but only from time T onwards, the intercept value we would obtain would be the value of Q. We would then use the above test statistic, replacing α by P and $\hat{\alpha}$ by the Q value. The procedure is then identical.

3.2   Confidence intervals of estimators

So far we have been solely concerned with point estimation. There may be research situations where we need to find some range of values within which we can be confident that the population parameter lies. This is the confidence interval of the parameter and can be calculated using the sample estimate of the parameter and the variance of the sampling distribution from which the estimate is obtained, along with a t value representing the degree of confidence we require.

3.2.1   Confidence interval for the regression slope

We have seen that the regression slope estimator is distributed normally with a mean of $\hat{\beta}$ and a variance of

$$\frac{\sigma^2}{\sum (X_i - \bar{X})^2}$$

In carrying out any two-tailed test we are trying to determine whether the value of the estimate obtained is so disparate from a value of zero that it goes beyond the critical values selected. Suppose we have selected

Diagram 3.6

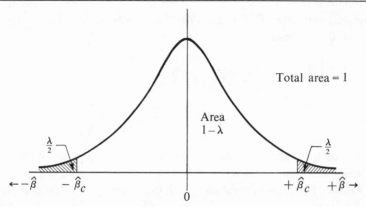

a two-tailed test and a significance level λ. As diagram 3.6 shows, the critical values of each tail will be at points beyond which $\lambda/2$ of the area under the curve exists. Our test statistic in this situation has been

$$t = \frac{\hat{\beta} - \beta}{\sqrt{\dfrac{\sigma^2}{\sum (X_i - \bar{X})^2}}}$$

but if we rearrange this equation so that β is on the left-hand side we obtain

$$\beta = \hat{\beta} - t \sqrt{\frac{\sigma^2}{\sum (X_i - \bar{X})^2}} \tag{3.13}$$

This quantity β actually represents the distance or interval between the center of the distribution and the critical value. If we consider both sides of the distribution, we will obtain an interval for the value of the population parameter β. Because the area between the two critical values is $1 - \lambda$ we can say that we are $1 - \lambda$ confident that β will lie between the two end points of the interval. This can be formally stated as

$$P\left[\hat{\beta} - t \sqrt{\frac{\sigma^2}{\sum (X_i - \bar{X})^2}} < \beta < \hat{\beta} \right.$$
$$\left. + t \sqrt{\frac{\sigma^2}{\sum (X_i - \bar{X})^2}} \right] = 1 - \lambda \tag{3.14}$$

As we have already seen, we cannot know the value of σ^2. We have to estimate this using the residuals from the estimated regression line and our best estimate of σ^2 is $\Sigma\, e_i^2/(n - K)$. Substituting this in the above equation we obtain

$$P\left[\hat{\beta} - t\sqrt{\frac{\sum e_i^2}{(n - K)\sum(X_i - \bar{X})^2}} < \beta < \hat{\beta}\right.$$

$$\left. + t\sqrt{\frac{\sum e_i^2}{(n - K)\sum(X_i - \bar{X})^2}}\right] = 1 - \lambda \quad (3.15)$$

Let us find the 95 percent confidence intervals for our money-spent–vote-increase example. When $1 - \lambda = 0.95$, $\lambda/2 = 0.025$. The critical t value for 16 degrees of freedom is 2.12. Our confidence interval is given by

$$P\left[0.002992 - (2.12)\sqrt{\frac{0.8513}{(18 - 2)(52,500,000)}} < \beta < 0.002996\right.$$

$$\left. + (2.12)\sqrt{\frac{0.8513}{(18 - 2)(52,500,000)}}\right] = 0.95$$

$$P[0.002929 < \beta < 0.003064] = 0.95$$

We are 95 percent confident that our population parameter lies between the values 0.002929 and 0.003064.

It is possible to procure such confidence limits for any regression parameter and with any specified confidence.

3.2.2   Confidence interval for the intercept

In an argument similar to the one given in Section 3.2.1, we can develop the following confidence intervals for the intercept:

$$P\left[\hat{\alpha} - t\sqrt{\frac{\sum e_i^2}{n(n - K)} \cdot \frac{\sum X_i^2}{\sum(X_i - \bar{X})^2}} < \alpha < \hat{\alpha}\right.$$

$$\left. + t\sqrt{\frac{\sum e_i^2}{n(n - K)} \cdot \frac{\sum X_i^2}{\sum(X_i - \bar{X})^2}}\right] = 1 - \lambda \quad (3.16)$$

And if we plug the data from our example into this relationship, we obtain

$$P[-0.064 - (2.12)(0.124) < \alpha < -0.064 + (2.12)(0.124)] = 0.95$$

which reduces to

$$P[-0.327 < \alpha < 0.199] = 0.95$$

which tells us that we are 95 percent confident that the true value of α lies between the values -0.327 and 0.199. Thus a combination of our data and the confidence we require can enable us to determine various ranges for the values of our population parameters.

3.2.3 Confidence interval for $\hat{Y}_i$

It is useful to know that for each of the observations of the explanatory variable we can create a confidence limit for the corresponding estimated explained variable. In this way we consider the precision of the entire regression line obtained from the data. We are thus testing its efficacy as a surrogate of the population regression line. In this case we use the estimate of Y_i as our starting value and proceed to develop an interval around this value.

As in the previous confidence interval development, we need to determine the variance of the estimate. The variance of the $\hat{Y}_i$ is:[3]

$$\text{var}(Y_i) = \sigma^2 \left[\frac{1}{n} + \frac{(X_i - \bar{X})^2}{\sum (X_i - \bar{X})^2} \right] \tag{3.17}$$

Not having σ^2, we use its estimator. Thus the variance of $\hat{Y}_i$ is

$$\text{var}(\hat{Y}_i) = \frac{\sum e_i^2}{n - K} \left[\frac{1}{n} + \frac{(X_i - \bar{X})^2}{\sum (X_i - \bar{X})^2} \right] \tag{3.18}$$

Notice that the variance is related to a particular value of $\hat{Y}_i$ and is not constant over all $\hat{Y}_i$. On inspecting the equation, we see that the variance of $\hat{Y}_i$ is smallest where the explanatory variable X_i takes a value nearest its mean. This is because at this point the quantity $(X_i - \bar{X})^2$ is small and may even be zero. Here the variance of $\hat{Y}_i$ reduces to

$$\text{var}(\hat{Y}_i) = \frac{\sum e_i^2}{n(n - K)} \tag{3.19}$$

In developing the confidence interval for this estimate we also require the

---

3. Kmenta, *Elements of Econometrics*, p. 222. Notice that Kmenta uses a slightly different format but the equation is the same in reality.

standard error of the estimate. This is equal to the square root of the variance. Our confidence interval equation becomes:

$$P\left[\hat{Y}_i - t \sqrt{\frac{\sum e_i^2}{n-K}\left(\frac{1}{n} + \frac{(X_i - \bar{X})^2}{\sum (X_i - \bar{X})^2}\right)} < Y_i < \hat{Y}_i \right.$$

$$\left. + t \sqrt{\frac{\sum e_i^2}{n-K}\left(\frac{1}{n} + \frac{(X_i - \bar{X})^2}{\sum (X_i - \bar{X})^2}\right)}\right] = 1 - \lambda \quad (3.20)$$

Comparison with the previously developed confidence intervals will illustrate that the same ingredients are there. They are the estimate, some chosen t value, and the standard error of the estimate. However, this equation is slightly different from the other confidence-interval equations in that the variance of the estimator changes with each value of the explanatory variable. So too will the confidence intervals. An example will make this more clear. Let us use our money-spent–vote-increase example.

I shall demonstrate the whole calculation for the first observation and then just display the other values in a table. We shall again take a 95 percent confidence interval which delivers up to us, when amalgamated with 16 degrees of freedom, a critical t value of 2.12. From our first observation we see that $X_1 = 1000$. Using our estimating equation,

$$\hat{Y}_1 = -0.064 + 0.002996(1000) = 2.932$$

The standard error of the estimate when $X_1 = 1000$ is given by

$$S_{\hat{Y}_1} = \sqrt{\frac{0.8513}{18-2}\left[\frac{1}{18} + \frac{(1000 - 3500)^2}{52.5 \times 10^6}\right]} = +0.096$$

Our confidence interval for the estimate of the first observation is

$$P[2.932 - 2.12(0.096) < Y_1 < 2.932 + 2.12(0.096)]$$
$$= P[2.729 < Y_1 < 3.136] = 0.95$$

Thus we can say that we are 95 percent confident that the true value of Y_1 lies between 2.927 and 3.136. We can do this for each observation in turn and produce the results shown in table 3.1. Notice how the interval between top and bottom limit narrows as we move towards the mean of

Table 3.1

X_1	Y_1	S_{Y_1}	$2.12\,S_{Y_1}$	Confidence Limits Lower	Upper
1000	2.932	0.096	0.2035	2.729	3.136
1000	2.932	0.096	0.2035	2.729	3.136
1000	2.932	0.096	0.2035	2.729	3.136
2000	5.928	0.072	0.1526	5.775	6.081
2000	5.928	0.072	0.1526	5.775	6.081
2000	5.928	0.072	0.1526	5.775	6.081
3000	8.924	0.057	0.1208	8.803	9.045
3000	8.924	0.057	0.1208	8.803	9.045
3000	8.924	0.057	0.1208	8.803	9.045
4000	11.920	0.057	0.1208	11.799	12.041
4000	11.920	0.057	0.1208	11.799	12.041
4000	11,920	0.057	0.1208	11.799	12.041
5000	14.916	0.072	0.1526	14.763	15.067
5000	14.916	0.072	0.1526	14.763	15.067
5000	14.916	0.072	0.1526	14.763	15.067
6000	17.912	0.096	0.2035	17.709	18.116
6000	17.912	0.096	0.2035	17.709	18.116
6000	17.912	0.096	0.2035	17.709	18.116

Diagram 3.7

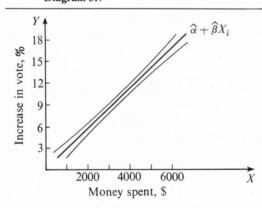

$\hat{\alpha} + \hat{\beta}X_i$

Increase in vote, %

Money spent, \$

X_1; as the values of X_i increase, the interval widens again. Diagram 3.7 shows this.

The variable width of the confidence interval is caused by possible errors in estimation of both the slope and the intercept. Diagram 3.8A shows the error band due to the slope. Since this slope must go through the intersection of the two means, possible estimates of the slope will rotate about that point. Diagram 3.8B indicates the error band due to our estimate of the intercept. This will be a parallel band. When the two errors are compounded we obtain, as in 3.8C, something akin to the diagram taken from the observations.

Diagram 3.8

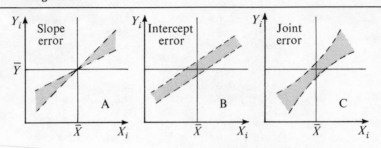

3.3 Prediction

Another feature of the generation of our regression line is that it allows us to predict outside the range of our empirical experience. Two kinds of predictions are possible: interpolation and extrapolation. In the former case we predict values within the range of observations of our explanatory variable, in the latter we predict outside this range. As we shall see when we develop our predictor, we are more confident within the range than outside it. This accords with common sense.

Not only do we require a prediction, however, but also some indication of the goodness of this prediction. This we obtain by creating confidence intervals around our prediction.

We want to predict a value of our explained variable given a value of our explanatory variable. Getting the prediction is relatively simple. We just substitute the value of the explanatory variable in our estimating equation and, behold, we obtain an estimate of the explained variable.

Developing the confidence interval of this estimate is somewhat more difficult. In this case, we are estimating one value and do not have the stabilizing influence of all the other possible values of Y. In effect this means that the variance of our estimate is larger. We have already seen that the variance of all possible estimated lines is given by the term

$$\text{var}\,(\hat{Y}_i) = \sigma^2 \left(\frac{1}{n} + \frac{(X_i - \bar{X})^2}{\sum (X_i - \bar{X})^2} \right) \tag{3.21}$$

But this is only the variance of the possible regression line estimates. As we know, around each of these lines there is a variance due to the error terms u_i (diagram 3.9). Fortunately, as diagram 3.9 suggests, variances are additive, and we know that the variance of the errors u_i is given by σ^2. This makes the total variance for our predictions $\hat{Y}_0$ equal to

$$\text{var}\,(\hat{Y}_0) = \sigma^2 \underbrace{\left[\frac{1}{n} + \frac{(X_i - \bar{X})^2}{\sum (X_i - \bar{X})^2} \right]}_{\text{line error variance}} + \underbrace{\sigma^2}_{\text{error variance}} \tag{3.22}$$

which becomes

$$\text{var}\,(\hat{Y}_0) = \sigma^2 \left[1 + \frac{1}{n} + \frac{(X_i - \bar{X})^2}{\sum (X_i - \bar{X})^2} \right] \tag{3.23}$$

Thus we are in a position to generate confidence limits for any prediction because we have the necessary ingredients—an estimate, a variance of this estimate, and a confidence level.

Diagram 3.9

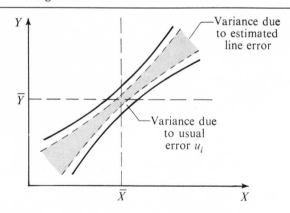

Let us take two examples, one from within the range of our observations and one from without. In the first example we desire to predict what value of Y is caused by a value of the explanatory variable of $4500 in our money-spent–vote-increase situation. Using our predictor equation, we get

$$\hat{Y}_0 = -0.064 + 0.002996(4500) = 13.418$$

The variance of this estimate is given by

$$\text{var}(\hat{Y}_0) = \frac{\sum e_i^2}{n-K}\left[1 + \frac{1}{n} + \frac{(X_i - \bar{X})^2}{\sum (X_i - \bar{X})^2}\right]$$

$$= \frac{0.8513}{(18-2)}\left[1 + \frac{1}{18} + \frac{10,000,000}{52,500,000}\right] = 1.075$$

$$\text{SD}(\hat{Y}_0) = 1.037$$

Thus our 95 percent confidence interval for this is

$$P[13.418 - 2.12(1.037) < \hat{Y}_0 < 13.418 + 2.12(1.037)] = 0.95$$

which gives us

$$P[11.22 < \hat{Y}_0 < 15.616] = 0.95$$

In our second example, we desire to predict what value of Y would be generated by a value of $10,000 in our explanatory variable. The estimated regression line produces

$$\hat{Y}_0 = -0.064 + 0.002996(10,000) = 29.898$$

and the variance of this estimate is given by

$$\text{var}(\hat{Y}_0) = \frac{0.8513}{(18-2)}\left[1 + \frac{1}{18} + \frac{(6,500)^2}{52,500,000}\right] = 1.86$$

$$\text{SD}(\hat{Y}_0) = 1.364$$

providing the 95 percent confidence interval

$$P[29.898 - 2.12(1.364) < \hat{Y}_0 < 29.898 + 2.12(1.364)] = 0.95$$

$$P[27.007 < \hat{Y}_0 < 32.789] = 0.95$$

The reader should notice that the confidence interval is far smaller within our range of observations than without. Thus our leap into the unknown is not made without providing some form of statistical safety net.

Further readings

Substantive

Most empirical studies using bivariate regression and correlation attempt some form of inference testing. The student can repair to these. Specific testing is found in:

Kirkpatrick, S. A. 1974. *Quantitative Analysis of Political Data.* Columbus, Ohio: Merrill.

Tufte, E. R. 1974. *Data Analysis for Politics and Policy.* Englewood Cliffs, N.J.: Prentice-Hall.

Tufte, E. R. 1970. *Quantitative Analysis of Social Problems.* Reading, Mass.: Addison-Wesley.

Statistical

Goldberger: pp. 163–70.
Johnston: pp. 26–35, 38–43.
Kmenta: pp. 216–29, 235–42.
Wonnacott and Wonnacott: pp. 23–34.

4
indicators of trouble in regression

We have gotten a little ahead of ourselves perhaps. By concentrating on testing hypotheses and developing predictions and confidence intervals, we have neglected a very important prior step—the examination of the regression-error distribution to determine whether any of the classical least-squares regression assumptions have been violated. Indeed, scrutiny of the regression estimates should be the last thing that we do. Prior to this there should be much search for contraventions of the assumptions.

We stated in Chapter 2 that there were five assumptions in regression. Four of these concerned the behavior of the error term and the fifth was related to the nonstochastic property of the explanatory variable. In this analysis we shall only concern ourselves with the assumptions regarding the error terms, which were:

(1) normality—the error term u_i is assumed to be normally distributed;
(2) zero mean—the mean value of u_i is zero;
(3) homoscedasticity—the variances of the error terms, σ_i^2, are constant, i.e.;
(4) no autocorrelation of the error terms—no error term should be related to a preceding or successive error term, i.e., the covariance of u_i and u_j is zero ($i \neq j$).

Violation of any or all of these assumptions may have effects upon the accuracy of both the estimates and the variances of the sampling distributions. Clearly, in view of what we learned in Chapter 3, this may have undesirable consequences for our inferential process. We ought, therefore, to develop mechanisms by which we can detect situations in which we may have violated the assumptions.

The most obvious mechanism is to examine the residuals of our regression. Remember that our best estimate of u_i is e_i, where e_i is the quantity $(Y_i - \hat{Y}_i)$. Examination of these residuals is comparatively easy since most computer regression programs print them out.

The most effective analysis of the residuals is visual. By plotting the residuals against variables included and excluded from the regression, we can quickly get some sense of possible violation. Mostly a visual analysis can be depended upon, but in some cases it can be inconclusive (rarely misleading!). In these cases we shall provide quantitative methods for determining delinquency.

We can divide the testing sequence into three sections. The first set of tests will inform us about the mean and normality of the residual distribution. The second section will investigate the existence of heteroscedasticity. The third section will look for possible violation of the autocorrelation assumption. In any regression analysis, all procedures should be carried out.

4.1   Testing assumptions about the residual distribution

Since the first assumption locates the mean of the residuals, the most obvious maneuver is to plot the residuals on some kind of point histogram. This plot will also allow us to consider the shape of the distribution to check visually for normality, which is our second assumption. Two such plots are shown in diagram 4.1. In the case on the left the residuals seem to be located near the zero point on the scale and the distribution looks normal. In contrast, the residual plot at right does not center around the zero point on the scale nor does the distribution look normal. We conclude in this case that we have violated the assumption, and consequences of this violation, which we shall discuss in the Chapter 5, will be reflected in our estimates.

The test for normality is far less clear cut. It is a property of the normal curve that within an interval of one standard deviation on either side of

Diagram 4.1

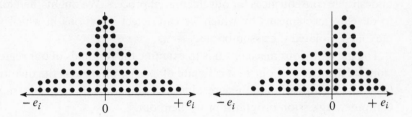

the mean will exist 68 percent of the total number of observations. Since we know both the mean and the standard deviation, it is relatively easy to check whether our curve is normal. The interval of two standard deviations on either side of the mean contains 95 percent of the total number of observations. This also can be checked. If the prescribed percentages of observations are found within these two intervals, this fact will add to our confidence that we have a normal distribution of residuals.

The difficulty with these tests is that with small sample sizes, it is not always possible to obtain reliable estimates of the standard error of the residuals. As the number of degrees of freedom gets small, so the estimated standard error becomes less stable. Nevertheless, between the visual and the quantitative tests it is usually possible to determine the characteristics of the residual distribution.

4.2   Tests for heteroscedasticity

Heteroscedasticity occurs when the variance of the residual changes with different values of the explanatory variable. Diagram 4.2 shows this effect. If we plot the residuals against each of the explanatory variables individually, heteroscedasticity is evidenced in patterns such as B, C, and D in diagram 4.3. Notice that the pictures show only a band of residuals rather than the plotted points themselves.

In case A we have homoscedasticity. All of the residuals are contained in a horizontal parallel band. In case B the width of the band increases as we move along the explanatory variable axis. We draw the inference that the explanatory variable is related to some component of the residual and is linked to this expansion in its variance. It is an indication that we have

Diagram 4.2

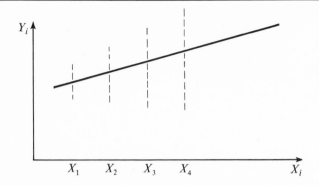

Diagram 4.3

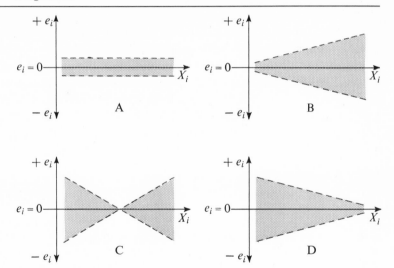

misspecified our regression model. This is also true of the situations depicted in C and D.

Notice that in all these cases the mean of residual distribution is approximately zero. There are more complex situations of heteroscedasticity where this may not be true (diagram 4.4). Although the width of the residual band is constant, it is the deviation from the zero that is used in the computation of the residual variance. In case 4.4A it may be that instead of using the value of the explanatory variable, the value of the

Diagram 4.4

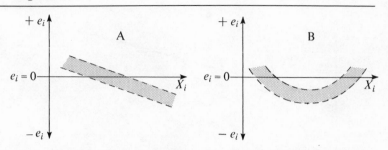

explanatory variable squared should have been used. In case 4.4B, perhaps other linear and quadratic values of the explanatory variable should have been used. Again, we conclude that we have misspecified the model. With heteroscedasticity a visual analysis is usually very informative and generally sufficient in pointing out any violations.

Another plot that gives similar information is that of the explained variable estimate $\hat{Y}_i$ against the residuals. As we know, $\hat{Y}_i$ is a linear aggregation of all the explanatory variables. Thus the plot will produce an aggregation of the plots of the explanatory variables. However, the patterns may not be so distinct since the effects of one variable might balance out the effects of another.

The detection of heteroscedasticity points to one or both of two weaknesses in our procedures:

(1) error in calculations, or
(2) misspecification of the model.

4.3   Tests for autocorrelation in residuals

Whereas heteroscedasticity is likely to occur in studies of a cross-sectional nature, autocorrelation occurs in longitudinal studies. It will occur in almost all longitudinal studies, being more pronounced if the interval between observations is short, less so if the time intervals are long. For instance, observations made every day will have severe autocorrelation problems, while observations collected annually may very well have less autocorrelation.

To detect autocorrelation in residuals graphically requires that the residuals be plotted over time. In this way it becomes possible to examine

Diagram 4.5

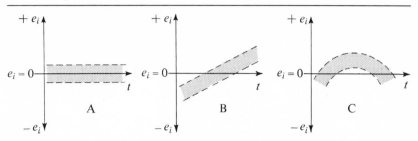

whether the residuals display any pattern over time. Should a pattern be discovered, autocorrelation is suspected. Diagram 4.5 shows some general patterns. In diagram 4.5A we may have a nonautocorrelated set of residuals. This is only a preliminary conclusion since we can have residuals that fit into a narrow horizontal band but that may still be autocorrelated, as in diagram 4.6. Diagram B obviously has autocorrelation. There is probably some linear term in time missing. In C, as time passes, the residuals change from negative to positive, then back to negative.

Diagram 4.6

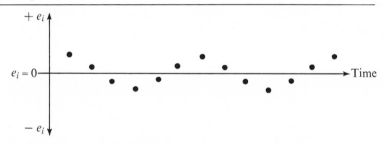

These are obvious autocorrelation patterns. A researcher would be very fortunate to obtain such distinct patterns; autocorrelation can remain undetected by graphical analysis. To surmount this difficulty we can provide a sophisticated statistical test for autocorrelation, called the Durbin-Watson test. Most computer regression programs compute this statistic automatically, although if not it is well worth the researcher's time to calculate the statistic manually.

The autocorrelation assumption requires that residuals not be correlated with one another. It more rigorously requires that no residuals be correlated, but for most purposes a first-order correlation prohibition is sufficient. Thus, if we plot e_{t-1} against e_t (where e_t is the residual at time t) we can obtain a visual estimate of their relationship. If they are not related, the points on the plot will be scattered with no pattern. Diagram 4.7A shows this. If they are related, some pattern will emerge as in Diagrams 4.7B and 4.7C. One way of detecting if they are linked is to determine the Pearson Product-moment correlation relating e_t to e_{t-1}. Let us call the correlation between successive residuals ρ. We want to test the following hypothesis:

$$H_0: \rho = 0 \quad H_A: \rho \neq 0$$

This can be done using the standard test for a hypothesis concerning the level of a correlation coefficient, but for small sample sizes there is likely to be some inaccuracy.

Diagram 4.7

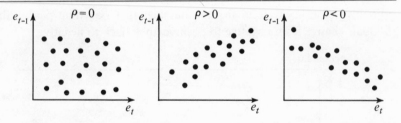

An alternative is the Durbin-Watson statistic, d, which can be calculated using

$$d = \frac{\displaystyle\sum_{t=2}^{n}(e_t - e_{t-1})^2}{\displaystyle\sum_{t=1}^{n} e_t^2} \qquad \text{where } e_t \text{ is the residual at time point } t \qquad (4.1)$$

And if we expand this equation, we get

$$d = \frac{\sum e_t^2}{\sum e_t^2} - 2\frac{\sum e_t e_{t-1}}{\sum e_t^2} + \frac{\sum e_{t-1}^2}{\sum e_t^2}$$

and letting

$$\frac{\sum e_t^2}{\sum e_t^2} = \frac{\sum e_{t-1}^2}{\sum e_t^2} = 1$$

approximately, we obtain

$$d \doteq 2\left(1 - \frac{\sum e_t e_{t-1}}{\sum e_t^2}\right)$$

where

$$\rho = \frac{\sum e_t e_{t-1}}{\sum e_t^2}$$

and thus the relationship reduces to an approximation

$$d \doteq 2(1 - \rho) \tag{4.2}$$

We can use this approximation to examine the implications of various d values. Notice that when $\rho = 0$, the Durbin-Watson statistic is approximately equal to 2. And as ρ becomes greater than 0, the value of d reduces until it finally arrives at 0 when $\rho = 1$, a perfect relationship between successive residuals. However, when ρ moves towards -1, the value of d moves higher than 2 and finally arrives at 4 when $\rho = -1$.

If we obtain a d value close to 2, we can accept the null-hypothesis and reject the alternative. There is no autocorrelation. However, as the value of d moves away from 2, we become more likely to reject the null-hypothesis and accept the alternative. The point at which we actually do reject the null-hypothesis is a function of the number of explanatory variables in the regression model K and the sample size n. There are tables for various combinations of numbers of explanatory variables and sample size.

Table 4.1

| Sample size n | Significance level 0.05 d_U and d_L one-tailed | | | |
| | $K = 2$ | | $K = 3$ | |
	d_L	d_U	d_L	d_U
15	1.08	1.36	0.95	1.54
16	1.10	1.37	0.98	1.54
17	1.13	1.38	1.02	1.5

Table 4.1 shows a small portion of them. I shall use this table for explanatory purposes. Note that we consider more than one explanatory variable here for the first time. This is to avoid having to go through the whole argument again when we deal with multivariate analysis in Part Two. The table provides—for a specific significance level ($p = 0.05$, a

specified number of variables K, and sample size n—two critical values called d_U and d_L, which are upper and lower limits respectively. With these critical values the decision table shown in diagram 4.8 can be constructed. Notice that when the correlation between successive residuals is low the value of d is close to 2. When the correlation gets positively high, d reduces to 0. And when the correlation gets negatively high, the value of d is close to 4. Also notice that the decision table is symmetrical around the value of 2. Although the Durbin-Watson tables only provide two critical points, we can use these twice to provide critical limits both above and below the d value of 2. Since we know the end points of the decision table, it is relatively simple to calculate the various critical values to the right of the $d = 2$ point.

Diagram 4.8

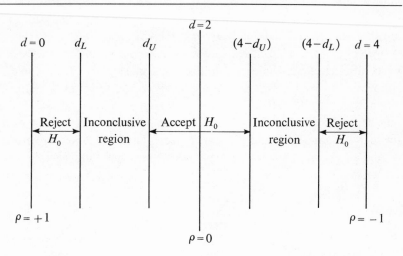

Use of the decision table is quite straightforward. If the d value obtained falls in the central two areas, the null-hypothesis is accepted. If the value falls in the two extreme end areas, we reject the null-hypothesis and accept the alternative that autocorrelation is present. If the value falls anywhere else, we conclude nothing about the hypothesis. This is the major difficulty with the Durbin-Watson statistic. There are two inconclusive regions in the decision map. Should a value fall in either of these, we can say nothing about the presence of autocorrelation of the residuals. An increase in sample size reduces this inconclusive region, but a larger sample is not always possible. In such situations the safest policy is to assume that there is autocorrelation and use the techniques provided in the next chapter to take care of the difficulty.

Using the money-spent–vote-increase example and calculating the d value, we obtain 2.51. Our sample size is 18 and the number of explanatory variables is 1. Using a 0.05 significance level, two-tailed, we produce the decision table shown in diagram 4.9. With this research situation, if we obtain a d value of 1.03 or less we reject H_0, as we do with a d value of 2.97 or more. If the value falls between 1.26 and 2.74 we accept the null-hypothesis. We conclude nothing if the value falls anywhere else.

Diagram 4.9

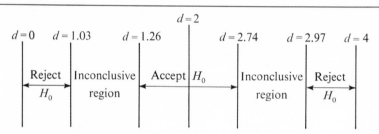

Since our value is 2.51, we accept the null-hypothesis that there is no autocorrelation of our residuals.

One final cautionary point. Durbin-Watson should not be used in situations where the explained variable lagged is used as one of the explanatory variables.

4.4   Some other considerations

Of the five classical least-squares assumptions, we have so far only presented techniques for detecting the violation of the four concerned with the error term. This is for the simple reason that these are more difficult to detect. As for detection of violations of the nonstochastic explanatory variable assumption, that is quite simple. If an experiment was carried out with various values of the explanatory variables being controlled, then we have probably not violated this assumption. Otherwise, we probably have.

In political science we violate this particular assumption in almost every research situation. The nonexperimental nature of the area of study forces this upon us.

Although we have concentrated on plots of the residuals to highlight violation of our assumptions, other benefits can be obtained. Any graphic representation of the residuals will emphasize values which are so deviant from other values that we can consider them to be outliers. Outliers can instruct us in many ways. First, an outlier can be caused by an error in the data collection. Obviously, a check on the data will indicate whether there has been an error. If there has, it can be corrected.

Second, by definition, the residuals point to those observations which are not summarized by the regression law. We can then examine each of these particular points to discover whether there is any theoretical reason why it does not conform to the regression law. This will give insight into variables that have been left out of the original regression model.

Finally, let me re-emphasize the importance of examining the residuals from regression analysis. Before the researcher can confidently exploit the various regression estimates, there has to be confidence that in obtaining the estimates there have been no violations of the classical assumptions. The easiest way to accomplish this is examination of the residuals. Visually and statistically we can check for assumption violations. But more than this, examination of the residuals may reveal limitations in our regression model and perhaps provide clues about deficiencies.

Unsophisticated visual analysis of residuals is the *first* operation the sophisticated politometrician will carry out.

Further readings

Substantive

Unfortunately, few political scientists who have employed politometric methods have actively reported analysis of the residuals of their regression. Several are listed here.

Cnudde, C. F. 1972. "Theories of Political Development and the Assumptions of Statistical Models: An Evaluation of Two Models." *Comparative Political Studies* 5: 131.

Davis, O. T., M. A. H. Dempster, and A. Wildavsky. 1966. "A Theory of the Budgetary Process." *Am. Pol. Sci. Review* 60: 529.

Kramer, G. H. 1971. "Short-Term Fluctuations in U.S. Voting Behavior, 1896–1964." *Am. Pol. Sci. Review* 65: 131.

McCally, S. P. 1966. "The Governor and His Legislative Party." *Am. Pol. Sci. Review* 60: 923.
Tufte, E. R. 1974. *Data Analysis for Politics and Policy.* Englewood Cliffs, N.J.: Prentice-Hall.
Tufte, E. R. 1969. "Improving Data Analysis in Political Science." *World Politics* 21: 641.

Statistical

Particularly:
Draper, N. R., and H. Smith. 1967. *Applied Regression Analysis.* New York: Wiley.

Also:
Kmenta: pp. 247–306.
Wonnacott and Wonnacott: pp. 132–48.

5

overcoming problems in regression

Violation of the five assumptions in classical regression can lead to inaccuracies in our estimating procedure. Both the estimators and the sampling distribution of the estimators are vulnerable to such inaccuracies. Reality, unfortunately, does not conform to the regression assumptions, with the result that any researcher is constantly confronted with difficult research situations. Whereas the statistician can develop techniques for making inferences from samples to populations in the rather abstract environment of mathematical statistics, the pragmatic politometrician has to develop technical maneuvers which enable him to cope with typical research situations which are rarely classical. This is the distinct difference between the statistician and the politometrician.

In Chapter 4 we developed methods for detecting violation of the regression assumptions. In this chapter we shall do two things. First, we shall discuss the effects of violations of the assumptions on our estimating procedure. Second, we shall develop strategies for overcoming the difficulties presented by violation of the assumptions. The reader should understand at the outset that not every situation can be dealt with adequately. In most cases we can construct serviceable alternative procedures but sometimes there will be nothing that can be done. Knowledge of what can and, more importantly, what cannot be done is the essence of the careful scientist.

What follows is in four sections. The first will look at violations of the assumptions of the error distribution. The second section will consider the difficulties encountered in heteroscedastic situations. A third section will confront the problem of autocorrelation of the error terms. A final section will provide particular comfort for those of us in areas of nonexperimental research. It will deal with stochastic explanatory variables.

5.1   Violation of the error distribution assumptions

The assumptions are that the mean of the error distribution is zero and that the distribution is normal. Violation of either of these presents no major problem for the researcher.

We can consider the normality provision first. If we drop this assumption, the least-square regression estimators are still BLUE. That is, they:

(1) are unbiased,
(2) are a linear function of the observations,
(3) have the smallest variance of all linear estimators.

We cannot say that the estimators are efficient because we cannot determine whether there are nonlinear estimators that have a smaller variance. But this is not important since we always finally have to speak of relative efficiency rather than absolute efficiency. We can never tell whether, in the universe of estimators, any one has the smallest variance.

The importance of the normality assumption lies in the hypothesis-testing sequences. All of the testing procedures, F, t, and Z, can only be strictly justified within the context of a normal curve. However, the tests have the quality of robustness. That is, they withstand reasonable deviations from the ideal. We can take comfort in this. If there are inaccuracies in the testing procedure, they will be critical when the test statistic is close to the boundaries of the acceptance and rejection regions. At times when the situation produces this proximity, the researcher should be cautious about inferences.

If the mean of the error distribution is not zero but some other value, say, v, then there will be a bias in the estimate of α. The calculated estimate will contain the unbiased estimate of α plus the value of v. This occurs because the regression procedure tries to locate the line through the mean of the distribution of error terms. Diagram 5.1 shows how the bias is produced.

Diagram 5.1

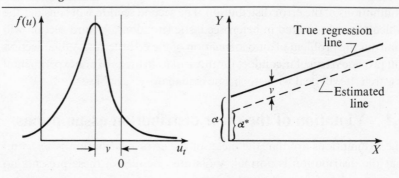

5.2   The problem of heteroscedasticity

We have seen that in classical regression the variance of the error distribution is required to be constant over all observations of the explanatory variable. This being the case, we can collapse all of the variance into one summarizing variance σ^2.

$$\sigma_1{}^2 = \sigma_2{}^2 = \sigma_3{}^2 = \cdots = \sigma^2$$

We calculate the variances of the sampling distribution of $\hat{\alpha}$ and $\hat{\beta}$ using equations (3.3) and (3.4).

$$\text{var}\,(\beta) = \frac{\sigma^2}{\sum (X_i - \bar{X})^2} \quad \text{var}\,(\alpha) = \frac{\sigma^2}{n} \left(\frac{\sum X_i{}^2}{\sum (X_i - \bar{X})^2} \right)$$

Obviously, if we cannot assume that all $\sigma_i{}^2 = \sigma^2$ then we will import bias into the above calculations. So our first conclusion is that heteroscedasticity introduces bias into sampling distribution variances. The bias will be reflected in our hypothesis-testing procedures because they depend on the sampling distribution variance. We might possibly err in our inferences.

Another salient point is that σ^2 does not feature in calculation equations for the estimates $\hat{\alpha}$ and $\hat{\beta}$. These equations are

$$\hat{\alpha} = \bar{Y} - \hat{\beta}\bar{X}$$

$$\hat{\beta} = \frac{\sum XY - \dfrac{\sum X \sum Y}{n}}{\sum X^2 - \dfrac{(\sum X)^2}{n}}$$

We may conclude from this that the estimators of the regression equation are unaffected by the changing variance of the error terms. They will remain unbiased; they are also still a linear function of the observations; and it can be shown that the estimators are still consistent.

But the bias introduced into the sampling distribution variance punishes the efficiency characteristic. This means that the least-squares estimators do not have the smallest variance of all possible estimators.

The real crippling feature, however, is that the variance of the sampling distributions of the regression estimators is biased, so *all* of the testing procedures developed in Chapter 3 are inapplicable.

Diagram 5.2 demonstrates why the variance of the sampling distribution is biased. It shows that the data points at high values of the explanatory variable have more of an effect on the variance of the sampling distribution than those at lower values of the explanatory variable. This, however, gives us some clue as to methods by which we can overcome this difficulty. If we can in some way weight these larger variances so that they have less effect on the sampling variance, we may regain reasonably accurate estimates of the sampling variance.

Diagram 5.2

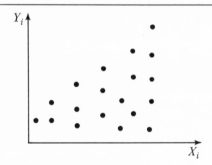

5.2.1 Heteroscedasticity caused by one explanatory variable

The selection of the weighting component is crucial. We need some weight which changes the variance at each observation of the explanatory variable such that each error variance makes an identical contribution to the overall sampling variance of the estimators.

The most obvious initial step is to link the variance of the error term to the value of the explanatory variable. Intuitively, as X_i changes so does the error variance. We can begin by stating the following relationship:

$$\sigma_i^2 = X_i^2 \sigma^2 \tag{5.1}$$

Verbally, this prescribes that the variance of the error term at observation i is a function of the explanatory variable X_i. The value of the explanatory variable is squared for mathematical reasons. This relationship can be rearranged to produce

$$\frac{\sigma_i^2}{X_i^2} = \sigma^2 \tag{5.2}$$

The quantity on the right-hand side is constant and is the term that we require for obtaining accurate estimates of the variance of the sampling distribution. Thus by transforming all of the data using the multiple $1/X_i$ we can get accurate estimates of the variance σ^2. This is called weighted least-squares (WLS), and we retain all the favorable characteristics of our estimators. The whole procedure consists of the following steps:

(1) Carry out ordinary least-squares analysis (OLS) on raw observations. This will produce residuals which can be plotted against all of the explanatory variables individually. (To avoid having to argue the same procedure for more than one explanatory variable, we shall consider situations with more than one explanatory variable here rather than in Part Two after multiple regression.) We can visually determine which, if any, of the explanatory variables is the culprit creating heteroscedasticity.

(2) If only one of the explanatory variables is causing heteroscedasticity, then transform the data by dividing all the variables (both explained and explanatory) by the value of the culprit. Then run WLS on the transformed regression equation

$$\frac{Y_i}{X_i} = \alpha \frac{1}{X_i} + \beta \tag{5.3}$$

Notice that the slope in the original equation now becomes the intercept in the WLS equation, and the intercept in the OLS becomes the slope in the WLS equation. It is important to remember this so that one knows which is which. The variance of the intercept coefficient from the WLS equation is really the unbiased variance of the slope coefficient for the required slope in the first equation. Similarly, the slope coefficient from the WLS regression is really our estimate of the intercept in the OLS equation. The whole process again using equation form is

(1) regress original model,
(2) look for heteroscedasticity by plotting the residuals against each of the explanatory variables in turn,
(3) if one of the explanatory variables is thought to cause hetero-scedasticity, divide throughout by this variable producing the following equation:

$$\frac{Y_i}{X_i} = \alpha \frac{1}{X_i} + \beta$$

This can be rearranged to

$$Y_i^* = \alpha^* + \beta^* X_i^*$$

where

$$Y_i^* = \frac{Y_i}{X_i}, \quad \alpha^* = \beta, \quad \beta^* = \alpha \quad \text{and} \quad X_i^* = \frac{1}{X_i}$$

Carry out OLS.

(4) rewrite the original equation using the estimates of the param-eters and the parameter variances computed from the weighted least-squares regression.

An example would seem to be called for at this point (diagram 5.3). The student should understand that it really does not matter how many

Diagram 5.3

Scattergram plot of raw data

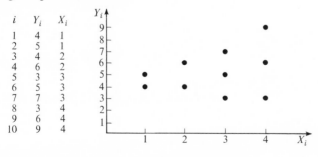

i	Y_i	X_i
1	4	1
2	5	1
3	4	2
4	6	2
5	3	3
6	5	3
7	7	3
8	3	4
9	6	4
10	9	4

Diagram 5.4

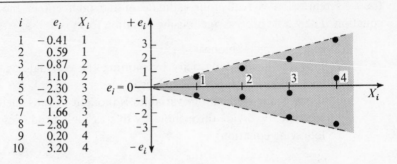

i	e_i	X_i
1	−0.41	1
2	0.59	1
3	−0.87	2
4	1.10	2
5	−2.30	3
6	−0.33	3
7	1.66	3
8	−2.80	4
9	0.20	4
10	3.20	4

explanatory variables there are, as long as only one of them is causing the heteroscedasticity. Regressing the data in diagram 5.3 using OLS, we obtain

$$\hat{Y}_i = \underset{(1.59)}{3.9504} + \underset{(0.554)}{0.463X_i} \qquad \text{Pearson } r = 0.286$$

Using $e_i = Y_i - \hat{Y}_i$, we can obtain the residuals. The plot of the residuals against the increasing value of the explanatory variable detects hetero-scedasticity (diagram 5.4). We transform the data in the problem by dividing by the explanatory variable, as in table 5.1. When we regress these data, we obtain:

$$\hat{Y}_i^* = \underset{(0.389)}{0.406} + \underset{(0.707)}{4.093X_i^*}$$

Table 5.1

i	$Y_i/X_i = Y_i^*$	$1/X_i = X_i^*$
1	4.00	1.00
2	5.00	1.00
3	2.00	0.50
4	3.00	0.50
5	1.00	0.33
6	1.67	0.33
7	2.33	0.33
8	0.75	0.25
9	1.50	0.25
10	2.25	0.25

To show the effect of the weighting process, let us again produce and plot the residuals for this WLS regression equation (diagram 5.5). The second regression on the weighted values of the variable was carried out under the optimal condition of homoscedasticity.

Diagram 5.5

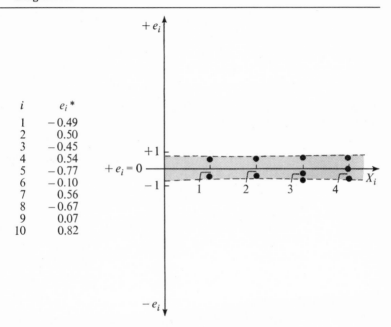

i	$e_i{}^*$
1	−0.49
2	0.50
3	−0.45
4	0.54
5	−0.77
6	−0.10
7	0.56
8	−0.67
9	0.07
10	0.82

Notice how the variance of the sampling distributions has changed. The variance of $\hat{\alpha}$ from the first equation was 2.54, but after the second regression this became 0.50. The variance of $\hat{\beta}$ from the first regression was 0.307, but after the second it became 0.151.

Finally, notice that the actual estimates of the parameters themselves have hardly changed; $\hat{\alpha}$ went from 3.9504 to 4.093 and $\hat{\beta}$ went from 0.463 to 0.406. This is evidence for our assertion about the unbiased nature of the estimates despite heteroscedasticity.

Thus, the final estimating equation is:

$$\hat{Y}_i = 4.093 + 0.406X_i \quad r = 0.178$$
$$\quad\;\;(0.707) \quad (0.389)$$

Notice that the Pearson correlation coefficient has been reduced to 0.178 from 0.286. We conclude that OLS gives us a better data fit and always will over every maneuver we employ in this chapter. The researcher, however, can now proceed to carry out any or all of the various tests described in Chapter 3.

5.2.2 Heteroscedasticity caused by more than one explanatory variable

If we believe, after analyzing the residuals visually, that more than one of the explanatory variables is causing heteroscedasticity, instead of weighting the regression equation by one explanatory culprit we weight using a linear combination of them all. The most obvious combination is $\hat{Y}_i$ because $\hat{Y}_i$ is the combination of all the explanatory variables weighted by their respective regression slopes.

Suppose we have a model with two explanatory variables, X_1 and X_2. The statistical model will be

$$Y_i = \alpha + \beta_1 X_{i1} + \beta_2 X_{i2} + u_i$$

The estimator model for this will be

$$\hat{Y}_i = \hat{\alpha} + \hat{\beta}_1 X_{i1} + \hat{\beta}_2 X_{i2}$$

We see that $\hat{Y}_i$ is a linear combination of X_1 and X_2. This being so, we can use $\hat{Y}_i$ as the weight in our weighted least-squares procedure when we feel that there is more than one culprit causing heteroscedasticity. The WLS regression thus becomes

$$\frac{Y_i}{\hat{Y}_i} = \alpha \frac{1}{\hat{Y}_i} + \beta_1 \frac{X_{i1}}{\hat{Y}_i} + \beta_2 \frac{X_{i2}}{\hat{Y}_i}$$

Notice that this equation has no constant term in it. This means that the regression is forced through the origin. This weighting technique is virtually the same as the previous one with the exception that it exploits a linear combination of all of the explanatory variables.

The whole process consists of four stages.

(1) Regress the original model.

$$Y_i = \alpha + \beta_1 X_{i1} + \beta_2 X_{i2}$$

(2) Look for heteroscedasticity by plotting the residuals against all of explanatory variables in turn.

So far this procedure is exactly the same as the previous one; it differs from here or when heteroscedasticity is found to be linked to more than one explanatory variable.

(3) If more than one of the explanatory variables is associated with the heteroscedasticity, divide the regression equation throughout by the $\hat{Y}_i$ obtained from the original OLS. This produces the following regression equation:

$$\frac{Y_i}{\hat{Y}_i} = \alpha \frac{1}{\hat{Y}_i} + \beta_1 \frac{X_{i1}}{\hat{Y}_i} + \beta_2 \frac{X_{i2}}{\hat{Y}_i}$$

which can be arranged to

$$Y_i^* = \beta_0 X_{i0}^* + \beta_1 X_{i1}^* + \beta_2 X_{i2}^*$$

where

$$Y_i^* = \frac{Y_i}{\hat{Y}_i}, \quad \beta_0 = \alpha, \quad X_{i0}^* = \frac{1}{\hat{Y}_i}, \quad X_{i1}^* = \frac{X_{i1}}{\hat{Y}_i}, \quad X_{i2}^* = \frac{X_{i2}}{\hat{Y}_i}$$

(4) Rewrite the original equation using the estimates obtained from the WLS regression.

Although this technique is designed to cope with situations where there is more than one explanatory variable, we shall stick with the example used in the previous weighted least-squares. This will not only demonstrate the technique but will give us some comparability between the two techniques.

From the previous situation we know that the estimating line for the sample data is

$$\hat{Y}_i = 3.9504 + 0.463 X_i$$
$$\quad\; (1.59) \qquad (0.554)$$

We also know that there is severe heteroscedasticity in the residuals. To cope with this we use WLS and the weight $\hat{Y}_i$. The values of $\hat{Y}_i$ for this example are given in table 5.2. When we transform the data by dividing

Table 5.2

i	$\hat{Y}_i$	i	$\hat{Y}_i$
1	4.413	6	5.339
2	4.413	7	5.339
3	4.876	8	5.802
4	4.876	9	5.802
5	5.339	10	5.802

through by this weighting, we obtain the following regression equation:

$$Y_i^* = \beta_0 X_{i0}^* + \beta_1 X_{i1}^*$$

The data for this are in table 5.3. We regress these data, remembering to force the regression through the origin. On doing this we obtain the following estimated line:

$$\hat{Y}_i^* = \underset{(1.349)}{4.005} X_{i0}^* + \underset{(0.496)}{0.442} X_{i1}^*$$

and when we put these back into the original equation we get

$$\hat{Y}_i = \underset{(1.349)}{4.005} + \underset{(0.496)}{0.442} X_{i1}$$

Table 5.3

i	Y_i^*	X_{i0}^*	X_{i1}^*	i	Y_i^*	X_{i0}^*	X_{i1}^*
1	0.91	0.23	0.23	6	0.94	0.19	0.56
2	1.13	0.23	0.23	7	1.31	0.19	0.56
3	0.82	0.21	0.41	8	0.52	0.17	0.69
4	1.23	0.21	0.41	9	1.03	0.17	0.69
5	0.56	0.19	0.56	10	1.55	0.17	0.69

5.2.3   Generalized weighted least-squares

The two procedures developed thus far in WLS are particular cases of a general approach. More generally we can consider that the variance of σ_i^2 is related to some variable Z_i. Thus we can put down the equation

$$\sigma_i^2 = \sigma^2 Z_i^\delta \tag{5.4}$$

The δ gives more generality to the relationship, allowing it a power function. By way of example we see that in our first example of WLS we allowed $Z_i = X_i$ and $\delta = 2$, thereby producing $\sigma_i^2 = \sigma^2 X_i^2$, the equation referred to at that time. In the second example where more than one explanatory variable is thought to cause heteroscedasticity, we let $Z_i = \hat{Y}_i$ and $\delta = 2$, producing

$$\sigma_i^2 = \sigma^2(\hat{Y}_i)^2$$

These are the two most common arrangements and are based upon assumptions concerning the cause of heteroscedasticity. Obviously there are countless other assumptions that could be made which would alter the weight used in WLS. And finally, when $\delta = 0$, the equation reduces to $\sigma_i^2 = \sigma^2$ or a homoscedastic situation.

5.3   The problem of autocorrelation of the error terms

While heteroscedasticity is likely to occur in cross-sectional data and not in time-series data, autocorrelation behaves in exactly the opposite manner. Time-series data are almost certain to contain autocorrelated error terms. The severity of the problem will depend somewhat upon the time interval between data points. As this interval increases, the severity of the problem will be reduced.

Autocorrelation of the error terms violates the fourth classical least-squares assumption. It will be remembered that this required that no error distribution, u_i, be correlated with any other error distribution, u_j. While this assumption requires that no two error distributions be related, we shall only be dealing with the simplest case of all, where there may be interdependence between consecutive error distributions. More formally, we shall be dealing with situations in which the following is true:

$$\text{cov}\,(u_t u_{t-1}) \neq 0 \tag{5.5}$$

Notice here that the subscript of the error term has been changed to t from i. This is to remind the student that we deal here with time-series data.

Why should autocorrelation pose a threat to our estimating procedure? We can answer this by considering the situation in which there is a positive correlation between consecutive error terms. A positive error term at one point in time will probably be followed by another positive error term. This, in turn, will probably be followed by a further positive error term, and so on. Should a negative error term be generated, however, this would be followed by another negative error term. Diagram 5.6 shows this in sketch A. The trace of error terms is consistently positive compared to the

Diagram 5.6

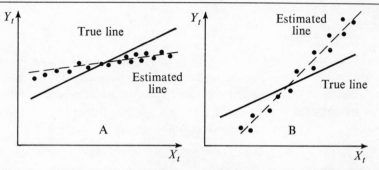

true line for the first half of the time period. After that some disturbance in the system sends the error negative and it remains negative. Notice, however, that the OLS procedure will try to locate a regression line through the data points. The estimated line will be different from the true line. But we have chosen a particular pattern. The error terms could just have easily started negative and then gone positive, as in sketch B, in which case the estimated regression line would have been biased the other way. Which of these two situations we get is left to chance. Thus, we can say that our estimator is unbiased, since the chance of its being one value or another is random and so on average it will be on the true value. But we also see that the variance of the slope and the intercept will be inaccurate because of the possible wide range of values of the estimated line around the true line. Knowledge that the variances of our sampling distribution are inaccurate renders inapplicable all of the inferential testing procedures developed in Chapter 3.

To be more specific, autocorrelation of the error terms produces unbiased, inefficient but still consistent estimators. So we will still have some of the attractive properties in our parameter estimates. Unfortunately, when we look at the variances of the sampling distributions of the estimators we find that they are biased and inefficient. In short, totally useless.

5.3.1   First-order autocorrelation
In suggesting that successive error terms are correlated we are stipulating the following relationship:

$$u_t = \rho u_{t-1} + v_t \qquad \text{where } u_t = \text{error at time } t$$

$$u_{t-1} = \text{error at time } t-1 \qquad (5.6)$$

In this equation the term v_t represents some other random error component which has all the properties that we require of error terms. Thus when $\rho = 0$, the whole thing reduces to $u_t = v_t$, with u_t possessing the qualities of v_t. When $\rho = 0$, u_t is a random error term with no autocorrelation. It can be shown that ρ is equivalent to the Pearson product-moment correlation between u_t and u_{t-1}.

5.3.2   Generalized differences to combat autoregression
We want our error terms to have the properties of the random error term v_t. To do this we rearrange equation (5.6) to

$$u_t - \rho u_{t-1} = v_t \tag{5.7}$$

This gives us a clue to the data transformation that we require to get us out of the autocorrelation difficulty. Let us consider our standard bivariate model,

$$Y_t = \alpha + \beta X_t + u_t$$

On finding that u_t is correlated to u_{t-1}, our transformation should be such that we obtain v_t as our error term. The following generalized differences equation manages this:

$$(Y_t - \rho Y_{t-1}) = \alpha(1 - \rho) + \beta(X_t - \rho X_{t-1}) + (u_t - \rho u_{t-1}) \tag{5.8}$$

This equation can be rewritten in the more familiar

$$Y_t^* = \alpha^* + \beta X_t^* + v_t \quad \text{where } Y_t^* = (Y_t - \rho Y_{t-1})$$
$$\alpha^* = \alpha(1 - \rho)$$
$$X_t^* = (X_t - \rho X_{t-1}) \tag{5.9}$$

This regression model does not violate the autocorrelation assumption.

Our generalized difference method for obtaining better estimates consists of four steps. In the first step ρ is estimated. The second step of the process is to transform the raw data according to equation (5.8). The third step is to carry out ordinary least-squares on the transformed data. The estimates thus produced have the qualities that we require for the testing sequences of inference testing, confidence-interval development, and the generation of predictions. However, because of the transformation equation, the estimate of the intercept α and its sampling variance must be detransformed, the fourth step. This is done quite easily by the following relationships:

$$\alpha = \frac{\alpha^*}{1 - \rho} \quad \text{var}(\alpha) = \frac{\text{var}(\alpha^*)}{(1 - \rho)^2}$$

5.3.3 Four ways of estimating ρ

The first step in exploiting the generalized differences method for combatting autoregression is providing an estimate of ρ for use in the second step. There are many ways of doing this, and we shall consider four.

The coefficient ρ measures the association between the distributions u_t and u_{t-1}. As usual we do not have either of these two distributions and are forced to use the estimate of u_t, the residual e_t. The most obvious manner of obtaining an estimate of the association between successive residuals is to calculate it using a revised version of the regular Pearson product-moment correlation coefficient. The revised version is

$$\rho = \frac{\sum e_t e_{t-1}}{\sum e_{t-1}^2} \quad (t = 2, 3, 4, \ldots, n) \tag{5.10}$$

Before we can use the residuals we have to generate them, and this is done by carrying out ordinary least-squares on the original data. We shall do this for the data in diagram 5.7. We can now obtain the correlation value ρ, using equation (5.10). Note that since we are lagging by one time period, we lose one observation, reducing the sample size to 19. If we carry out this calculation for the residuals in the example, we obtain an estimate for $\hat{\rho} = 0.165$.

A second method for obtaining an estimate for ρ concerns the use of equation (5.11)

$$e_t = \rho e_{t-1} + v_t \tag{5.11}$$

In this method we regress e_{t-1} on e_t using ordinary least-squares, making sure to force the regression through the origin because there is no constant in the equation. Most computers have a facility for forcing the regression through the origin. The slope coefficient obtained is a least-squares estimate of ρ. Using the data in the example, we obtain a second estimate of $\hat{\rho} = 0.187$. The two methods used thus far have generated two estimates of ρ. The estimates are relatively close to one another.

Diagram 5.7

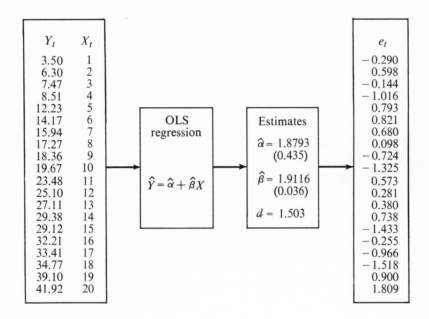

Y_t	X_t			e_t
3.50	1			−0.290
6.30	2			0.598
7.47	3			−0.144
8.51	4			−1.016
12.23	5			0.793
14.17	6	OLS	Estimates	0.821
15.94	7	regression		0.680
17.27	8		$\hat{\alpha} = 1.8793$	0.098
18.36	9		(0.435)	−0.724
19.67	10			−1.325
23.48	11	$\hat{Y} = \hat{\alpha} + \hat{\beta}X$	$\hat{\beta} = 1.9116$	0.573
25.10	12		(0.036)	0.281
27.11	13			0.380
29.38	14		$d = 1.503$	0.738
29.12	15			−1.433
32.21	16			−0.255
33.41	17			−0.966
34.77	18			−1.518
39.10	19			0.900
41.92	20			1.809

The third technique for getting an estimate of ρ is to exploit the Durbin-Watson statistic equation. This equation stipulates that

$$d \doteq 2(1 - \rho) \quad \text{or} \quad \rho \doteq 1 - \frac{d}{2}$$

In our example the value for d is 1.503. Our estimate of $\hat{\rho}$ is 0.25. This estimate is a little disparate from the others.

The fourth and final technique we shall discuss is another Durbin method. With this technique we take the generalized differences equation and rearrange it. The generalized difference equation is

$$Y_t - \rho Y_{t-1} = \alpha(1 - \rho) + \beta(X_t - \rho X_{t-1}) + v_t$$

which we arrange to

$$Y_t = \alpha(1 - \rho) + \rho Y_{t-1} + \beta X_t - \rho \beta X_{t-1} + v_t \qquad (5.12)$$

This expression is treated as a regular regression with three explanatory variables, Y_{t-1}, X_t, and X_{t-1}. The slope coefficient that we obtain for the variable Y_{t-1} is another estimator of ρ. Using this technique on the data, we obtain a fourth estimate of 0.181 for $\hat{\rho}$.

The estimates we have obtained for $\hat{\rho}$ are 0.165, 0.187, 0.25, and 0.181. We now use the generalized difference equation to get estimators of the intercept and slope coefficients and their sampling distribution variance. This is the second stage in this process.

We have four estimates of ρ. Let us take one of them and complete all of the calculations. We can then put down the results for the insertion of the other three estimates of ρ and leave all the calculation steps out, for they will be the same as in the first case.

The generalized difference equation will be

$$Y_t^* = \alpha^* + \beta X_t^* + v_t \quad \text{where } Y_t^* = Y_t - 0.165Y_{t-1}$$
$$\alpha^* = \alpha(1 - 0.165)$$
$$X_t^* = X_t - 0.165X_{t-1}$$

Thus in transforming the data to get Y_t^* and X_t^* we obtain the information in diagram 5.8.

To obtain our estimate of $\hat{\alpha}$ and its sampling variance, we have to detransform as below:

$$\hat{\alpha} = \frac{\hat{\alpha}^*}{1 - \rho} \quad \text{and} \quad \text{var}(\hat{\alpha}) = \frac{\text{var}(\hat{\alpha}^*)}{(1 - \rho)^2}$$

This gives us in this case $\hat{\alpha} = 1.91$ and $\text{var}(\hat{\alpha}) = 0.36$. This completes the four-stage procedure and our final estimating equation using generalized differences, and an estimate of $\rho = 0.165$ is

$$\hat{Y_t} = 1.91 + 1.9118X_t$$
$$\begin{array}{cc} (0.6) & (0.048) \end{array}$$

Notice also how the maneuver has produced a better Durbin-Watson value of 1.65, which is closer to 2 (no autocorrelation) than the 1.503 obtained in the original least-squares calculation.

Carrying out the above procedure for all of the four different estimates of ρ produces the regression estimated lines shown in table 5.4. Each should be compared with our original estimated line, which is at the bottom of the table.

Diagram 5.8

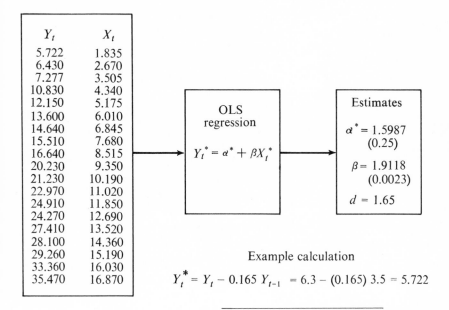

Y_t	X_t
5.722	1.835
6.430	2.670
7.277	3.505
10.830	4.340
12.150	5.175
13.600	6.010
14.640	6.845
15.510	7.680
16.640	8.515
20.230	9.350
21.230	10.190
22.970	11.020
24.910	11.850
24.270	12.690
27.410	13.520
28.100	14.360
29.260	15.190
33.360	16.030
35.470	16.870

OLS regression

$$Y_t^* = \alpha^* + \beta X_t^*$$

Estimates

$$\alpha^* = 1.5987$$
$$(0.25)$$

$$\beta = 1.9118$$
$$(0.0023)$$

$$d = 1.65$$

Example calculation

$$Y_t^* = Y_t - 0.165\, Y_{t-1} = 6.3 - (0.165)\,3.5 = 5.722$$

Example calculation

$$X_t^* = X_t - 0.165\, X_{t-1} = 2 - (0.165) = 1.835$$

Table 5.4

	Generalized Differences	
$\rho = 0.165$	$\hat{Y}_t = 1.91 + 1.9118X_t$ (0.60) (0.048)	$d = 1.65$
$\rho = 0.187$	$\hat{Y}_t = 1.91 + 1.9126X_t$ (0.62) (0.05)	$d = 1.678$
$\rho = 0.250$	$\hat{Y}_t = 1.89 + 1.9154X_t$ (0.67) (0.053)	$d = 1.755$
$\rho = 0.181$	$\hat{Y}_t = 1.91 + 1.9124X_t$ (0.61) (0.049)	$d = 1.671$
OLS	$\hat{Y}_t = 1.87 + 1.9116X_t$ (0.435) (0.036)	$d = 1.503$

All of the estimates for $\hat{\alpha}$ and $\hat{\beta}$ are roughly the same. This supports the assertion that autocorrelation does not introduce bias into the estimating procedure. Notice, however, that although the variances of the sampling distribution for the four estimating lines using generalized differences are fairly constant, the variances of the sampling distribution in the OLS equation are quite inaccurate. Such inaccuracy might lead to wrong inferences. For any particular significance level, the null-hypothesis with the estimates derived here from OLS have more chance of being rejected than in any of the generalized differences estimates.

Finally, the Durbin-Watson statistic is closer to 2 in all of the generalized differences equations.

In summary, let us go once more through the technique for overcoming problems of autocorrelation of the error terms. If autocorrelation is suspected:

Step 1. Obtain an estimate of ρ by any one of the four methods described.

Step 2. Use this estimate of ρ to transform the data according to the generalized differences equation.

Step 3. Carry out OLS regression using the transformed data.

Step 4. Detransform the intercept coefficient and variance.

The wisest and most conservative procedure is to develop all four of the estimates as above. Employ each of them in the generalized difference equation to obtain four different regression lines. To choose between these lines, select the one with the largest variance. Thus for any significance level the researcher will be more likely to accept the null-hypothesis.

5.4   Stochastic explanatory variable

The final violation concerns the nature of the explanatory variable, X_i. It will be remembered that X_i is fixed by the experimenter and there should be equal probabilities of getting different values of X_i. There is little that we can do in situations where this assumption is violated. Because of this, all that will be done here is to declare the effects of this violation upon the estimators and their variances. In this way we can at least make judgments about the inaccuracies in our procedures. This is one step better than disregarding the inaccuracies.

When the explanatory variable is stochastic, we find that there is a probability distribution of values of this variable. This means that some values of the variable will be more evident than other values. For instance, certain temperatures are more likely than others in Chicago. During the winter we are unlikely to see 70°F. just as we are unlikely to see temperatures of $-10°$F. in the summer. Temperature is thus exhibiting stochastic properties. It is reasonable to state that in all nonexperimental situations the explanatory variable will be stochastic. And in political science, which is almost exclusively nonexperimental, we almost always violate this assumption. But it is not quite as bad as it sounds. In many situations this violation will have no effect on any of our procedures. We can distinguish three situations.

(1) X and u are completely uncorrelated,
(2) X and u are contemporaneously uncorrelated,
(3) X and u are not uncorrelated nor contemporaneously uncorrelated.

Each case will be discussed to point out the effects on the estimator properties. When the explanatory variable and the error term are uncorrelated, the estimators still retain their property of unbiasedness. The estimators are also efficient. So we see that a relaxation of this assumption does not lead to a loss of the attractive properties of the estimates when the explanatory variable and the error term are uncorrelated.

When we consider the variance of the least-squares estimators we find that they are the same as when the explanatory variable is nonstochastic. In summary, it can be said that, in this particular situation of stochastic explanatory variable uncorrelated with the error terms, our testing procedures are still perfectly viable.

In the second of the three situations the explanatory variable is contemporaneously uncorrelated with the error term. This may occur when one of the explanatory variables is the explained variable lagged. For example,

$$Y_t = \alpha + \beta Y_{t-1} + u_t$$

Given this model we also know that

$$Y_{t-1} = \alpha + \beta Y_{t-2} + u_{t-1}$$

This tells us that Y_{t-1} is correlated with u_{t-1} and since Y_{t-1} acts as a explanatory variable in the first model, it will be the case that Y_t is also correlated with u_{t-1}. But unless u_t is correlated with u_{t-1} there is no way that Y_{t-1} can be correlated with u_{t-1}. Thus, while Y_{t-1} is correlated with u_{t-1}, it is not contemporaneously correlated with u_t. In this situation it can be shown that the classical testing procedures are only valid asymptotically. That is, as long as we have fairly large samples we can still estimate accurately and infer correctly from our sample to the population.

In the final situation, where the explanatory variable and the error terms are neither uncorrelated nor contemporaneously uncorrelated, we are in extreme difficulty. As diagram 5.9 indicates, the estimates are biased. The error is positive for low values of X and negative for high values of X. Thus X is correlated with the error term. The estimated line will be inaccurate. Not only are the estimates biased, but they are inconsistent. Indeed, it is a useless situation as it stands. Situations like this occur frequently when we have simultaneous equations, interacting equations. Alternative methods of estimation have been developed for these situations and these will be discussed when we deal with multivariate, multiequation models.

Diagram 5.9

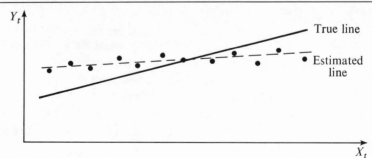

Further Readings

Substantive

Unhappily, I again have to report that very few practicing politometricians have employed any of the recovery techniques described in this chapter. Here is one that I located.

Cnudde, C. F. 1972. "Theories of Political Development and the Assumptions of Statistical Models: An Evaluation of Two Models." *Comparative Political Studies* 5: 131.

Statistical

Particularly:
Draper, N. R., and H. Smith. 1967. *Applied Regression Analysis.* New York: Wiley.

Also:
Kmenta: chapter 8.
Wonnacott and Wonnacott: chapter 6.

Part Two

This second part of the book is devoted to carrying out all of the previously described operations on models that have more than one explanatory variable.

To assist in this, Chapter 6 provides the student with a basic understanding of matrix algebra. This will also prove useful in later parts of the book. Chapters 7 and 8 cover estimation and hypothesis testing, respectively, using matrix algebra.

The final chapter in Part Two is a catch-all chapter dealing with regression models which are related to the normal regression format. Binary, nonlinear, and stepwise regression models are discussed along with models which have variables from different time periods.

6

matrix algebra: visual arithmetic

It is now convenient to introduce the reader to a different way of doing arithmetic. We are moving into the research situations which are characterized by many variables in each equation. For each of these variables we will have a distribution of observations and we can represent them in patterns. Let there be four explanatory variables which we shall denote as X_{i1}, X_{i2}, X_{i3}, and X_{i4}. The first subscript refers to the observation number and the second refers to the variable. Thus X_{32} represents the third observation on the second explanatory variable. The data matrix will be:

		Explanatory variables			
Case 1	Y_1	X_{11}	X_{12}	X_{13}	X_{14}
Case 2	Y_2	X_{21}	X_{22}	X_{23}	X_{24}
Case 3	Y_3	X_{31}	X_{32}	X_{33}	X_{34}
.	.	.	.	.	.
Case n	Y_n	X_{n1}	X_{n2}	X_{n3}	X_{n4}

The data are set out visually, and with little difficulty we can locate any particular observation on any variable. Our data are in a numerical pattern. Operations on this pattern can be considered as pattern arithmetic. So as well as retaining an arithmetical power, we also facilitate a visual display of the numbers. We can see which numbers from the original pattern are used to produce numbers in other patterns. This will give us more insight into the various arithmetic maneuvers and their consequences. This form of arithmetic is called matrix algebra. It allows us to operate upon such patterns of numbers according to sets of rules, many of which are augmented variations of simple arithmetic rules.

This topic is introduced here, rather than in an appendix where it might very well go in note form, because it is powerful in dealing with data matrices. The operations involved are relatively simple, and any reader who finishes the examples will have an adequate knowledge for the rest of the text. This form of arithmetic will be foreign to many political scientists, but do persevere! It will be worth the effort in terms of grasping the arguments which follow; but more than this, mastery of matrix algebra can provide much intellectual and visual pleasure.

6.1　Definitions

6.1.1

A *matrix* is a set of numbers arranged in an array of rows and columns. Each of the numbers in the matrix is called an *element* of the matrix. Each matrix will have m rows and n columns. The *order* of the matrix is $m \times n$. In the matrix below, which we denote by $\mathbf{X}$, there are three rows and four columns. The matrix thus has an order of 3×4. It should be noted that we shall use boldface capital letters for all matrix symbols while each element of the matrix will be a small letter with subscripts, e.g., x_{21}. Thus we can locate each number according to the number of its row i and the number of its column j. The element x_{21} is in the second row of the matrix and the first column. More generally each element can be represented by x_{ij}.

$$\mathbf{X} = \begin{bmatrix} 3 & 3 & 6 & 2 \\ 4 & 0.5 & 5 & -3 \\ 2 & 6 & 8 & 8 \end{bmatrix} \text{— Element } x_{34}$$

Matrix of order 3×4

The elements of the matrix can be any number from minus to plus infinity.

6.1.2

There are some matrices which either have one row or one column. These are more correctly specified as *vectors*. A vector with only one row is called a *row vector*. A row vector is a $1 \times n$ matrix. A vector with only one column is called a *column vector*. A column vector is an $m \times 1$ matrix. In what follows, vectors will be represented by boldface letters. The examples of a row vector and column vector are taken from the matrix above.

$$\mathbf{a} = \begin{bmatrix} 3 \\ 0.5 \\ 6 \end{bmatrix} \qquad \mathbf{X} = [2 \quad 6 \quad 8 \quad 8]$$

Column vector (3×1) — Row vector (1×4)

6.1.3

A *square matrix* is one with the number of rows equal to the number of columns. A *diagonal matrix* has numbers other than zero only in the diagonal going from top left to bottom right of a square matrix. A diagonal matrix is shown below.

$$\mathbf{A} = \begin{bmatrix} 6 & 0 & 0 \\ 0 & 13 & 0 \\ 0 & 0 & 12 \end{bmatrix}$$

3×3 Diagonal matrix

6.1.4

The special case of the diagonal matrix is the *unity or identity matrix*. In this matrix all the top-left to lower-right diagonals are equal to one, and all of the off-diagonals are zero.

$$\mathbf{I} = \begin{bmatrix} 1 & 0 & 0 & 0 \\ 0 & 1 & 0 & 0 \\ 0 & 0 & 1 & 0 \\ 0 & 0 & 0 & 1 \end{bmatrix}$$

Identity matrix

6.1.5

A zero matrix is one in which all of the elements are zero.

$$\mathbf{0} = \begin{bmatrix} 0 & 0 & 0 \\ 0 & 0 & 0 \\ 0 & 0 & 0 \end{bmatrix}$$

Zero matrix

6.1.6

The *trace* of a matrix is the summation of all the diagonal elements of a square matrix. Every square matrix will have a trace whether or not the off-diagonals are all zero.

$$P = \begin{bmatrix} 3 & 4 & 7 \\ 6 & 1 & 2 \\ 9 & 8 & 2 \end{bmatrix} \quad \begin{array}{l} \text{tr}(P) = 6 \\ \text{Note: tr}(I) = m \text{ or } n \end{array}$$

6.1.7

A *scalar* is a 1 × 1 matrix, that is, a single number.

6.1.8

Matrix **R** is equal to matrix **S**, if and only if every element of **R** is the same as every element of **S**.

$$R = \begin{bmatrix} 2 & 8 & -4 \\ 1 & 3 & 2 \\ 9 & 6 & 11 \end{bmatrix} \quad S = \begin{bmatrix} 2 & 8 & -4 \\ 1 & 3 & 2 \\ 9 & 6 & 11 \end{bmatrix}$$

6.2   Basic matrix operations

6.2.1   Addition

Matrices can be added if and only if they have the same order. The resulting matrix will also be of that order.

$$A = \begin{bmatrix} 7 & 2 \\ 3 & 1 \end{bmatrix}; \quad B = \begin{bmatrix} 5 & 2 \\ -3 & 5 \end{bmatrix}$$

$$A + B = \begin{bmatrix} 7+5 & 2+2 \\ 3+(-3) & 1+5 \end{bmatrix} = \begin{bmatrix} 12 & 4 \\ 0 & 6 \end{bmatrix}$$

6.2.2   Subtraction

Matrices can be subtracted from one another if and only if they have the same order. The resulting matrix will also be of that order.

$$A - B = \begin{bmatrix} 7-5 & 2-2 \\ 3-(-3) & 1-5 \end{bmatrix} = \begin{bmatrix} 2 & 0 \\ 6 & -4 \end{bmatrix}$$

6.2.3   Multiplication

Multiplication requires more care than either addition or subtraction, for only matrices of certain orders may be multiplied. Multiplication between two matrices **A** and **B** can only take place if the number of *columns* of **A** equals the number of *rows* of **B**. Such matrices are considered conformable. The resulting matrix **C** will have an order equal to the number of rows of **A** by the number of columns of **B**.

The actual multiplication is rather difficult to explain verbally. To obtain the first element in the first row of the final matrix **C**, multiply the first element in the first row of matrix **A** by the first element in the first column of matrix **B**; multiply the second element in the first row of matrix **A** by the second element in the first column of matrix **B**; and so on. Then add these products. For the rest of the elements in the first row of matrix **C**, repeat procedure: add the products of each element in the first row of **A** multiplied by the corresponding element in successive columns of **B**. For the remaining rows of matrix **C**, repeat, adding the products of each element in successive rows of **A** times the corresponding element in the columns of **B**, across all columns.

Let us take the following example. The requirement is to multiply **A** by **B**.

$$\mathbf{A} = \begin{bmatrix} 2 & 5 \\ 3 & 6 \\ 4 & 8 \end{bmatrix} \quad \mathbf{B} = \begin{bmatrix} 4 & 9 & 2 & 1 \\ 8 & 3 & 4 & 6 \end{bmatrix}$$

Notice that the number of columns of **A** is equal to the number of rows of **B**, i.e., they are conformable. We can thus multiply these together.

$$\mathbf{C} = \begin{bmatrix} 2 \times 4 + 5 \times 8 & 2 \times 9 + 5 \times 3 & 2 \times 2 + 5 \times 4 & 2 \times 1 + 5 \times 6 \\ 3 \times 4 + 6 \times 8 & 3 \times 9 + 6 \times 3 & 3 \times 2 + 6 \times 4 & 3 \times 1 + 6 \times 6 \\ 4 \times 4 + 8 \times 8 & 4 \times 9 + 8 \times 3 & 4 \times 2 + 8 \times 4 & 4 \times 1 + 8 \times 6 \end{bmatrix}$$

from row 3 of **A**   from column 1 of **B**

$$\mathbf{C} = \begin{bmatrix} 48 & 33 & 24 & 32 \\ 60 & 45 & 30 & 39 \\ 80 & 60 & 40 & 52 \end{bmatrix}$$

Notice that the order of the new matrix is 3 × 4. This is the number of rows in **A** by the number of columns in **B**.

Notice also that although we can produce **C** = **AB** we *cannot* produce the product **BA** because **B** has four columns and **A** has three rows. From this we can see that **AB** may not equal **BA**. And, indeed, in our example, **BA** does not exist. We can distinguish between pre- and post-multiplication. In multiplying **B** by **A**, pre-multiplication would be **AB**, while post-multiplication would be **BA**.

We can, however, generate matrices which exist when there is both pre- and post-multiplication. Notice that **AB** and **BA** are not equal. Only in very special cases will **AB** = **BA**.

$$\mathbf{A} = \begin{bmatrix} 6 & 4 \\ 1 & 3 \\ 2 & 1 \end{bmatrix} \quad \mathbf{B} = \begin{bmatrix} 4 & 0 & 3 \\ 1 & 2 & 1 \end{bmatrix}$$

$$\mathbf{AB} = \begin{bmatrix} 24+4 & 0+8 & 18+4 \\ 4+3 & 0+6 & 3+3 \\ 8+1 & 0+2 & 6+1 \end{bmatrix} \quad \mathbf{BA} = \begin{bmatrix} 24+0+6 & 16+0+3 \\ 6+2+2 & 4+6+1 \end{bmatrix}$$

$$\mathbf{AB} = \begin{bmatrix} 28 & 8 & 22 \\ 7 & 6 & 6 \\ 9 & 2 & 7 \end{bmatrix} \quad \mathbf{BA} = \begin{bmatrix} 30 & 19 \\ 10 & 11 \end{bmatrix}$$

Finally, an easy way to remember the rule regarding multiplication is *across and down*. The number of elements *across* in the first matrix must be equal to the number of elements *down* in the second matrix.

Scalar multiplication is the multiplication of each element in a matrix by some scalar λ. There are no particular rules to be offered here. Suppose we want to multiply matrix **A** by a scalar of 3.

$$\mathbf{A} = \begin{bmatrix} 2 & 5 \\ 3 & 6 \\ 4 & 8 \end{bmatrix}; \quad \lambda = 3 \quad \lambda\mathbf{A} = \begin{bmatrix} 3 \times 2 & 3 \times 5 \\ 3 \times 3 & 3 \times 6 \\ 3 \times 4 & 3 \times 8 \end{bmatrix} = \begin{bmatrix} 6 & 15 \\ 9 & 18 \\ 12 & 24 \end{bmatrix}$$

6.2.4 Division

Matrix division is a little more complex and will be discussed later.

Scalar division is relatively simple. Each of the elements in the matrix to be divided by the scalar is individually divided by the scalar. The order of the matrix so produced is the same as the initial matrix.

$$\mathbf{A} = \begin{bmatrix} 2 & 5 \\ 3 & 6 \\ 4 & 8 \end{bmatrix}; \quad \lambda = 3 \qquad \frac{\mathbf{A}}{\lambda} = \begin{bmatrix} \frac{2}{3} & \frac{5}{3} \\ \frac{3}{3} & \frac{6}{3} \\ \frac{4}{3} & \frac{8}{3} \end{bmatrix} = \begin{bmatrix} 0.67 & 1.67 \\ 1.00 & 2.00 \\ 1.33 & 2.67 \end{bmatrix}$$

6.2.5 Transposition

In any matrix $\mathbf{A}$ which has elements a_{ij}, the transpose is another matrix $\mathbf{A}'$ with elements a_{ji}. To transpose any matrix we take each element in the matrix, reverse the subscripts, and relocate the element in a new matrix. So what we do is interchange rows and columns. The first row of $\mathbf{A}$ becomes the first column of $\mathbf{A}'$. The second row of $\mathbf{A}$ becomes the second column of $\mathbf{A}'$ and so on. For example:

$$\mathbf{A} = \begin{bmatrix} 6 & 4 \\ 4 & 6 \\ 1 & 2 \\ 2 & 1 \end{bmatrix} \quad \mathbf{A}' = \begin{bmatrix} 6 & 4 & 1 & 2 \\ 4 & 6 & 2 & 1 \end{bmatrix}$$

and

$$\mathbf{X} = \begin{bmatrix} 6 & 2 & 4 \\ 4 & 8 & 3 \\ 6 & 9 & 1 \end{bmatrix} \quad \mathbf{X}' = \begin{bmatrix} 6 & 4 & 6 \\ 2 & 8 & 9 \\ 4 & 3 & 1 \end{bmatrix}$$

There are some relationships that may be useful. It is left to the reader to prove the relationships. This can be quite easily done using example matrices.

(1) $(\mathbf{A}')' = \mathbf{A}$
(2) $(\mathbf{A} + \mathbf{B})' = \mathbf{A}' + \mathbf{B}'$
(3) $(\mathbf{A} - \mathbf{B})' = \mathbf{A}' - \mathbf{B}'$
(4) $(\mathbf{AB})' = \mathbf{B}'\mathbf{A}'$
(5) $(\mathbf{ABC})' = \mathbf{C}'\mathbf{B}'\mathbf{A}'$
(6) $(\mathbf{A}'\mathbf{A})' = \mathbf{A}'(\mathbf{A}')' = \mathbf{A}'\mathbf{A}$ symmetric matrix
(7) $(\mathbf{AA}')' = (\mathbf{A}')'\mathbf{A}' = \mathbf{AA}'$ symmetric matrix

We have introduced in the last two relationships the idea of the *symmetric* matrix. This is a matrix where the transpose is equal to the original matrix. If $\mathbf{A} = \mathbf{A}'$ we have a symmetric matrix. For example:

$$\mathbf{A} = \begin{bmatrix} 1 & 2 & 3 & 4 \\ 2 & 2 & 6 & 8 \\ 3 & 6 & 3 & 2 \\ 4 & 8 & 2 & 4 \end{bmatrix} \quad \mathbf{A}' = \begin{bmatrix} 1 & 2 & 3 & 4 \\ 2 & 2 & 6 & 8 \\ 3 & 6 & 3 & 2 \\ 4 & 8 & 2 & 4 \end{bmatrix}$$

Obviously, all symmetric matrices will be square.

There are some special arrangements of matrices and their transposes which are particularly valuable for our politometric calculations. If $\mathbf{A}$ is an $m \times n$ matrix, then $\mathbf{A}'$ will be an $n \times m$ matrix. And if we multiply these together, that is produce $\mathbf{AA}'$ or $\mathbf{A}'\mathbf{A}$, we will end up with an $m \times m$ or $n \times n$ matrix, respectively. Notice that both of these matrices are square. For example:

$$\mathbf{A} = \begin{bmatrix} 2 & 3 & 4 \\ 5 & 6 & 8 \end{bmatrix} \quad \mathbf{A}' = \begin{bmatrix} 2 & 5 \\ 3 & 6 \\ 4 & 8 \end{bmatrix}$$

$$\mathbf{AA}' = \begin{bmatrix} 29 & 60 \\ 60 & 125 \end{bmatrix} \quad \mathbf{A}'\mathbf{A} = \begin{bmatrix} 29 & 36 & 48 \\ 36 & 45 & 60 \\ 48 & 60 & 80 \end{bmatrix}$$

Notice: (1) $\mathbf{A}'\mathbf{A} \neq \mathbf{AA}'$. This is important in politometrics!

(2) trace $(\mathbf{AA}')$ = trace $(\mathbf{A}'\mathbf{A})$.

(3) trace $(\mathbf{AA}')$ = trace $(\mathbf{A}'\mathbf{A})$ = sum of squares of the matrix $\mathbf{A}$.

6.2.6 Partitioned matrices

There are some situations where we are only interested in selected parts of the matrix. We can partition the matrix off so that only calculations on the relevant portion are carried out. In essence, all we do is break a large matrix down into smaller matrices according to our needs. The rules already described for full matrices are still operable on the smaller matrices. We can partition $\mathbf{A}$ into $\mathbf{A}_{11}$ and $\mathbf{A}_{12}$ as below.

$$\mathbf{A} = (\mathbf{A}_{11} \mid \mathbf{A}_{12})$$

and the transpose of $\mathbf{A}$, $\mathbf{A}'$ becomes

$$\mathbf{A}' = \frac{\mathbf{A}'_{11}}{\mathbf{A}'_{12}}$$

For example,

$$A = \begin{bmatrix} 1 & 2 & 2 & 1 \\ 2 & -1 & -2 & 3 \\ 1 & 4 & 5 & 2 \end{bmatrix} \quad A = \left[\begin{array}{cc|cc} 1 & 2 & 2 & 1 \\ 2 & -1 & -2 & 3 \\ 1 & 4 & 5 & 2 \end{array} \right]$$

where

$$A_{11} = \begin{bmatrix} 1 & 2 \\ 2 & -1 \\ 1 & 4 \end{bmatrix} \quad \text{and} \quad A_{12} = \begin{bmatrix} 2 & 1 \\ -2 & 3 \\ 5 & 2 \end{bmatrix}$$

Now,

$$A' = \left[\begin{array}{ccc} 1 & 2 & 1 \\ 2 & -1 & 4 \\ \hline 2 & -2 & 5 \\ 1 & 3 & 2 \end{array} \right] \quad \text{that is,} \quad A' = \begin{bmatrix} A'_{11} \\ A'_{12} \end{bmatrix}$$

All operations of addition, subtraction, and multiplication can be performed on partitioned matrices as long as the various partitions are *conformable*. That is, the same rules apply to the order of the partitioned matrices as to whole matrices. We shall use one example of addition, subtraction, and multiplication. Notice that each operation can only be carried out on conformable matrices.

Addition:

$$A = \left[\begin{array}{cc} 2 & 5 \\ \hline 3 & 6 \\ 4 & 8 \end{array} \right] \quad A_{11} = [2 \; 5] \quad A_{21} = \begin{bmatrix} 3 & 6 \\ 4 & 8 \end{bmatrix}$$

and

$$B = \left[\begin{array}{cc} 3 & 2 \\ \hline 1 & 4 \\ 6 & 1 \end{array} \right] \quad B_{11} = [3 \; 2] \quad B_{21} = \begin{bmatrix} 1 & 4 \\ 6 & 1 \end{bmatrix}$$

On addition,

$$A + B = \begin{bmatrix} A_{11} + B_{11} \\ A_{22} + B_{22} \end{bmatrix} = \left[\begin{array}{cc} 5 & 7 \\ \hline 4 & 10 \\ 10 & 9 \end{array} \right]$$

Notice how the order of $\mathbf{A}_{11}$ was equal to the order of $\mathbf{B}_{11}$ and that of $\mathbf{A}_{21}$ was equal to the order of $\mathbf{B}_{21}$.

Subtraction: Using the matrices from above we get

$$\mathbf{A} - \mathbf{B} = \begin{bmatrix} \mathbf{A}_{11} - \mathbf{B}_{11} \\ \hline \mathbf{A}_{22} - \mathbf{B}_{22} \end{bmatrix} = \begin{bmatrix} -1 & 3 \\ \hline 2 & 2 \\ -2 & 7 \end{bmatrix}$$

Multiplication: The most crucial point here is that not only must the whole matrices be multipliable but also the way in which they are partitioned should be in keeping with the *across and down* rule of multiplication.

Suppose $\mathbf{A}$ is of the order 3×2 and we partition the matrix thus:

$$\mathbf{A}_1 = \begin{bmatrix} a_{11} & a_{12} \\ a_{21} & a_{22} \\ \hline a_{31} & a_{32} \end{bmatrix} \quad \mathbf{A}_{11} = \begin{bmatrix} a_{11} & a_{12} \\ a_{21} & a_{22} \end{bmatrix} \quad \mathbf{A}_{21} = \begin{bmatrix} a_{31} & a_{32} \end{bmatrix}$$

Then in order to premultiply this we require some matrix $\mathbf{B}$ partitioned in the following manner:

$$\mathbf{B} = \begin{bmatrix} b_{11} & b_{12} & b_{13} & b_{14} \\ b_{21} & b_{22} & b_{23} & b_{24} \end{bmatrix}$$

$$\mathbf{B}_{11} = \begin{bmatrix} b_{11} & b_{12} \\ b_{21} & b_{22} \end{bmatrix} \quad \mathbf{B}_{12} = \begin{bmatrix} b_{13} & b_{14} \\ b_{23} & b_{24} \end{bmatrix}$$

Consequently when we multiply these together we shall obtain

$$\mathbf{AB} = \begin{bmatrix} \mathbf{A}_{11}\mathbf{B}_{11} & \mathbf{A}_{11}\mathbf{B}_{12} \\ \mathbf{A}_{21}\mathbf{B}_{11} & \mathbf{A}_{21}\mathbf{B}_{12} \end{bmatrix}$$

And we can see from this that not only must the whole matrix be conformable but also each partition in the multiplication must be conformable. That is, the number of columns in $\mathbf{A}_{11}$ must be equal to the number of rows in $\mathbf{B}_{11}$ and $\mathbf{B}_{12}$. The number of columns in $\mathbf{A}_{21}$ must be equal to the number of rows in $\mathbf{B}_{11}$ and $\mathbf{B}_{12}$. For example, we shall use a previous multiplication:

$$\mathbf{A} = \begin{bmatrix} 2 & 5 \\ 3 & 6 \\ 4 & 8 \end{bmatrix} \quad \mathbf{A}_{11} = \begin{bmatrix} 2 & 5 \\ 3 & 6 \end{bmatrix} \quad \mathbf{A}_{21} = \begin{bmatrix} 4 & 8 \end{bmatrix}$$

$$\mathbf{B} = \begin{bmatrix} 4 & 9 & 2 & 1 \\ 8 & 3 & 4 & 6 \end{bmatrix} \quad \mathbf{B}_{11} = \begin{bmatrix} 4 & 9 \\ 8 & 3 \end{bmatrix} \quad \mathbf{B}_{12} = \begin{bmatrix} 2 & 1 \\ 4 & 6 \end{bmatrix}$$

and

$$\mathbf{A}_{11}\mathbf{B}_{11} = \begin{bmatrix} 48 & 33 \\ 60 & 45 \end{bmatrix} \quad \mathbf{A}_{11}\mathbf{B}_{12} = \begin{bmatrix} 24 & 32 \\ 30 & 39 \end{bmatrix}$$

$$\mathbf{A}_{21}\mathbf{B}_{11} = \begin{bmatrix} 80 & 60 \end{bmatrix} \quad \mathbf{A}_{21}\mathbf{B}_{12} = \begin{bmatrix} 40 & 52 \end{bmatrix}$$

so

$$\mathbf{AB} = \begin{bmatrix} 48 & 33 & 24 & 32 \\ 60 & 45 & 30 & 39 \\ 80 & 60 & 40 & 52 \end{bmatrix}$$

The point of partitioning is that it allows concentration on any particular part of the matrix according to our interest.

6.3   Determinants

A *determinant* is a scalar value associated with each *square* matrix. The determinant of a matrix is obtained by multiplying elements of the matrix together according to a specified pattern. And the determinant is denoted by det $\mathbf{A}$ or $|\mathbf{A}|$. Let us consider the simplest case where we have a 2×2 matrix.

$$\begin{bmatrix} a_{11} & a_{12} \\ a_{21} & a_{22} \end{bmatrix}$$

In this case the determinant is given by the relationship $a_{11}a_{22} - a_{12}a_{21}$. That is, we have multiplied the top left-hand element by the bottom right-hand element and subtracted from this the top-right hand element multiplied by the bottom left-hand element. Notice that the diagonal going from top left to bottom right is positive and that going from top right to bottom left is negative. For example,

$$\mathbf{A} = \begin{bmatrix} 3 & 4 \\ 1 & 2 \end{bmatrix} \qquad \mathbf{X} = \begin{bmatrix} 3 & 4 \\ 2 & 2 \end{bmatrix}$$

$$|\mathbf{A}| = 3 \times 2 - 4 \times 1 = 2 \qquad |\mathbf{X}| = 3 \times 2 - 4 \times 2 = -2$$

We can now consider a 3×3 matrix. Such a matrix would have the general form

$$\mathbf{A} = \begin{bmatrix} a_{11} & a_{12} & a_{13} \\ a_{21} & a_{22} & a_{23} \\ a_{31} & a_{32} & a_{33} \end{bmatrix}$$

To calculate the determinant of this matrix, we take each of the elements in the top row individually and multiply them with the 2×2 matrix remaining when the elements in the same column as the multiplying element are deleted. In addition to this rather complex procedure we have to use some device for maintaining the correct sign in front of each term. This is achieved by raising -1 to the power of the sum of row number and column number. Thus when $i = 1$ and $j = 1$, that is, we are dealing with the top left-hand element a_{11}, the sign of the multiplication will be positive because $(-1)^{1+1} = 1$. The reader will quickly perceive that the pattern will be add, subtract, add. The determinant of a 3×3 matrix is calculated as below:

$$|\mathbf{A}| = (-1)^{1+1}a_{11}\underbrace{\begin{bmatrix} a_{22} & a_{23} \\ a_{32} & a_{33} \end{bmatrix}}_{\text{cofactor}} + (-1)^{1+2}a_{12}\underbrace{\begin{bmatrix} a_{21} & a_{23} \\ a_{31} & a_{33} \end{bmatrix}}_{\text{cofactor}}$$

$$+ (-1)^{1+3}a_{13}\underbrace{\begin{bmatrix} a_{21} & a_{22} \\ a_{31} & a_{32} \end{bmatrix}}_{\text{cofactor}}$$

$$= a_{11}[a_{22}a_{33} - a_{32}a_{23}] - a_{12}[a_{21}a_{33} - a_{31}a_{23}]$$
$$+ a_{13}[a_{21}a_{32} - a_{31}a_{22}]$$

The term

$$(-1)^{1+1}\begin{bmatrix} a_{22} & a_{23} \\ a_{32} & a_{33} \end{bmatrix}$$

is called the cofactor of a_{11}.

Notice the symmetry of the procedure. Clarification might very well be achieved by an example or two. After working through these examples go back to "We can now consider a $3 \times 3 \ldots$," and read the description of the procedure again.

$$\mathbf{X} = \begin{bmatrix} 3 & 4 & 3 \\ 2 & 9 & 2 \\ 6 & 2 & 1 \end{bmatrix}$$

$$|\mathbf{X}| = (-1)^{1+1} \times 3 \begin{bmatrix} 9 & 2 \\ 2 & 1 \end{bmatrix} + (-1)^{1+2} \times 4 \begin{bmatrix} 2 & 2 \\ 6 & 1 \end{bmatrix}$$

$$+ (-1)^{1+3} \times 3 \begin{bmatrix} 2 & 9 \\ 6 & 2 \end{bmatrix}$$

$$= 3(9 - 4) - 4(2 - 12) + 3(4 - 54)$$
$$= 3(5) - 4(-10) + 3(-50)$$

$$|\mathbf{X}| = -95$$

$$\mathbf{A} = \begin{bmatrix} 1 & 2 & 3 \\ 4 & 5 & 6 \\ 7 & 8 & 9 \end{bmatrix}$$

$$|\mathbf{A}| = (-1)^{1+1} \times 1 \begin{bmatrix} 5 & 6 \\ 8 & 9 \end{bmatrix} + (-1)^{1+2} \times 2 \begin{bmatrix} 4 & 6 \\ 7 & 9 \end{bmatrix}$$

$$+ (-1)^{1+3} \times 3 \begin{bmatrix} 4 & 5 \\ 7 & 8 \end{bmatrix}$$

$$= 1(45 - 48) - 2(36 - 42) + 3(32 - 35) = 0$$

This is a special kind of matrix. Any matrix that has a determinant of zero is called a *singular* matrix.

With higher-order matrices the same procedure is repeated except that one has many more steps to get the cofactors down to the final and usable 2 × 2 matrices. This can be a long and tedious process and is well left to the many computer programs that produce the determinant of any matrix. It is enough for the reader to appreciate what a determinant is and from whence it comes.

There are some very pleasing properties of determinants.

(1) If we interchange any two rows or any two columns, the sign of the determinant also changes. For example, we have seen that the determinant of the matrix below has a value of 2. The determinant of the other matrix, which is the first one with the columns switched, is −2.

$$A = \begin{bmatrix} 3 & 4 \\ 1 & 2 \end{bmatrix} \quad B = \begin{bmatrix} 4 & 3 \\ 2 & 1 \end{bmatrix}$$

$$|A| = 2 \qquad |B| = 4 - 6 = -2$$

(2) If a matrix **A** is multiplied by a scalar, then the determinant of the resulting matrix is equal to the determinant of **A** multiplied by λ^n where n is the order of the matrix. More formally,

$$|\lambda A| = \lambda^n |A|$$

For example,

$$A = \begin{bmatrix} 1 & 2 \\ 3 & 4 \end{bmatrix}; \quad \lambda = 3 \qquad \lambda A = \begin{bmatrix} 3 & 6 \\ 9 & 12 \end{bmatrix}$$

$$|A| = -2 \qquad\qquad |\lambda A| = 36 - 54 = -18 = 3^2(-2)$$
$$= \lambda^n |A|$$

(3) The value of a determinant of a matrix where two rows or columns are identical is zero. The matrix is singular. Also, if one row (or column) is linearly dependent upon another row (or column), that matrix will also be singular. This is a very important property in politometrics. Notice in the second example below that the first row is exactly half the second row.

$$X = \begin{bmatrix} 1 & 1 \\ 2 & 2 \end{bmatrix} \quad A = \begin{bmatrix} 1 & 2 \\ 2 & 4 \end{bmatrix}$$

$$|A| = 0 \qquad |A| = 0$$

(4) The determinant of any matrix is equal to the determinant of its transpose. Formally,

$$|A| = |A'|$$

For example,

$$A = \begin{bmatrix} 3 & 4 \\ 1 & 2 \end{bmatrix} \quad A' = \begin{bmatrix} 3 & 1 \\ 4 & 2 \end{bmatrix}$$

$$|A| = 2 \qquad |A'| = 6 - 4 = 2$$

(5) The determinant of the product of two matrices is equal to the product of the matrices' determinants. Formally,

$$|\mathbf{AB}| = |\mathbf{A}| \times |\mathbf{B}|$$

For example,

$$\mathbf{A} = \begin{bmatrix} 3 & 4 \\ 1 & 2 \end{bmatrix} \quad \mathbf{B} = \begin{bmatrix} 6 & 2 \\ 1 & 3 \end{bmatrix} \quad \mathbf{AB} = \begin{bmatrix} 22 & 18 \\ 8 & 8 \end{bmatrix}$$

$$|\mathbf{A}| = 2 \qquad |\mathbf{B}| = 16 \qquad |\mathbf{AB}| = 32$$

6.4 Matrix inversion

The inverse of an $n \times n$ matrix $\mathbf{A}^{-1}$ is another $n \times n$ matrix such that the product of the $\mathbf{A}^{-1}$ and the $\mathbf{A}$ is equal to an identity matrix for both pre- and post-multiplication. Formally,

$$\mathbf{AA}^{-1} = \mathbf{A}^{-1}\mathbf{A} = \mathbf{I} \quad \text{(the identity matrix)}$$

Suppose

$$\mathbf{X} = \begin{bmatrix} 6 & 4 \\ 3 & 1 \end{bmatrix} \quad \text{then} \quad \mathbf{X}^{-1} = \begin{bmatrix} -\frac{1}{6} & \frac{2}{3} \\ \frac{1}{2} & -1 \end{bmatrix}$$

When we multiply these together we get

$$\mathbf{XX}^{-1} = \begin{bmatrix} (6)(-\frac{1}{6}) + (4)(\frac{1}{2}) & (6)(\frac{2}{3}) + (4)(-1) \\ (3)(-\frac{1}{6}) + (1)(\frac{1}{2}) & (3)(\frac{2}{3}) + (1)(-1) \end{bmatrix} = \begin{bmatrix} 1 & 0 \\ 0 & 1 \end{bmatrix}$$

We can define the inverse of the matrix $\mathbf{A}$ as

$$\mathbf{A}^{-1} = \frac{1}{|\mathbf{A}|} \ (\text{adj } \mathbf{A})$$

Notice that if $\mathbf{A}$ is singular, that is, its determinant is zero, it becomes impossible to obtain the inverse of $\mathbf{A}$. This is an important result for politometrics. We shall define the adjoint of $\mathbf{A}$ (adj $\mathbf{A}$) as we go through the procedure.

The procedure for developing the inverse of a matrix consists of four steps and can only be achieved for square matrices. In the first example we require the inverse of the matrix $\mathbf{A}$:

$$\mathbf{A} = \begin{bmatrix} 6 & 4 \\ 3 & 1 \end{bmatrix}$$

> Step 1. Develop the determinant of $\mathbf{A}$. If the determinant of $\mathbf{A}$ is zero, the whole procedure should be terminated because $\mathbf{A}^{-1}$ will not exist.
>
> $$|\mathbf{A}| = 6 - 12 = -6$$

Since the matrix is nonsingular we proceed to step 2.

> Step 2. Produce the cofactor matrix. When we were considering the determinant we referred to the cofactor matrix. For this matrix $\mathbf{A}$ we obtain
>
> $$\text{Cofactor matrix} = \begin{bmatrix} (-1)^{1+1} \times 1 & (-1)^{1+2} \times 3 \\ (-1)^{2+1} \times 4 & (-1)^{2+2} \times 6 \end{bmatrix} = \begin{bmatrix} 1 & -3 \\ -4 & 6 \end{bmatrix}$$
>
> Step 3. Produce the adjoint of $\mathbf{A}$. The adjoint of $\mathbf{A}$ is equal to the cofactor matrix of $\mathbf{A}$ transposed. Thus adj $\mathbf{A}$ is given by
>
> $$\text{adj } \mathbf{A} = \begin{bmatrix} 1 & -4 \\ -3 & 6 \end{bmatrix}$$
>
> Step 4. Place necessary parts into the equation:
>
> $$\mathbf{A}^{-1} = \frac{1}{|\mathbf{A}|} [\text{adj } \mathbf{A}]$$

On doing this for our example, we obtain

$$\mathbf{A}^{-1} = \frac{1}{-6} \begin{bmatrix} 1 & -4 \\ -3 & 6 \end{bmatrix}$$

Thus $\mathbf{A}^{-1}$ is given by

$$\mathbf{A}^{-1} = \begin{bmatrix} -\frac{1}{6} & \frac{2}{3} \\ \frac{1}{2} & -1 \end{bmatrix}$$

and as we saw before, $\mathbf{A}\mathbf{A}^{-1} = 1$.

Let us go onto the larger matrix, a 3×3. For example,

$$\mathbf{A} = \begin{bmatrix} 4 & 5 & 1 \\ 1 & 1 & 1 \\ 3 & -2 & -1 \end{bmatrix}$$

Step 1. Obtain the determinant:

$$|\mathbf{A}| = 4(-1 + 2) - 5(-1 - 3) + 1(-2 + 3) = 19$$

Step 2. Develop the cofactor matrix:

$$\begin{bmatrix} (-1)^{1+1}\begin{bmatrix} 1 & 1 \\ -2 & -1 \end{bmatrix} & (-1)^{1+2}\begin{bmatrix} 1 & 1 \\ 3 & -1 \end{bmatrix} & (-1)^{1+3}\begin{bmatrix} 1 & 1 \\ 3 & -2 \end{bmatrix} \\ (-1)^{1+2}\begin{bmatrix} 5 & 1 \\ -2 & -1 \end{bmatrix} & (-1)^{1+3}\begin{bmatrix} 4 & 1 \\ 3 & -1 \end{bmatrix} & (-1)^{1+4}\begin{bmatrix} 4 & 5 \\ 3 & -2 \end{bmatrix} \\ (-1)^{1+3}\begin{bmatrix} 5 & 1 \\ 1 & 1 \end{bmatrix} & (-1)^{1+4}\begin{bmatrix} 4 & 1 \\ 1 & 1 \end{bmatrix} & (-1)^{1+5}\begin{bmatrix} 4 & 5 \\ 1 & 1 \end{bmatrix} \end{bmatrix}$$

which becomes

$$\begin{bmatrix} 1 & 4 & -5 \\ 3 & -7 & 23 \\ 4 & -3 & -1 \end{bmatrix}$$

Step 3. Obtain the transpose of this to get the adjoint:

$$\text{adj}(\mathbf{A}) = \begin{bmatrix} 1 & 3 & 4 \\ 4 & -7 & -3 \\ -5 & 23 & -1 \end{bmatrix}$$

Step 4. Produce the inverse:

$$\mathbf{A}^{-1} = \frac{1}{19}\begin{bmatrix} 1 & 3 & 4 \\ 4 & -7 & -3 \\ -5 & 23 & -1 \end{bmatrix} = \begin{bmatrix} \frac{1}{19} & \frac{3}{19} & \frac{4}{19} \\ \frac{4}{19} & -\frac{7}{19} & -\frac{3}{19} \\ -\frac{5}{19} & \frac{23}{19} & -\frac{1}{19} \end{bmatrix}$$

Obtaining the inverse of matrices larger than 3×3 is an extremely tedious process and best left to the computer. Nevertheless, it is important that the student understand the process since all of the matrix calculations for regression consist of determining inverses. Should these procedures go wrong, the student will be able to specify exactly why.

We shall now consider a very important matrix inversion. This is the $(\mathbf{X'X})$ matrix whose inversion is the basic matrix for all regression procedures. The $\mathbf{X}$ matrix is that which contains all of the observations of the explanatory variables plus a column of those for the intercept term. The model for multiple regression is

$$Y_i = \beta_0 + \beta_1 X_{i1} + \beta_2 X_{i2} + \cdots + \beta_{K-1} X_{i, K-1}$$

When we put each observation into this, we get

$$Y_1 = \beta_0 + \beta_1 X_{11} + \beta_2 X_{12} + \cdots + \beta_{K-1} X_{1, K-1}$$

$$Y_2 = \beta_0 + \beta_1 X_{21} + \beta_2 X_{22} + \cdots + \beta_{K-1} X_{2, K-1}$$

$$\cdot \qquad \cdot \qquad \cdot \qquad \cdot \qquad \cdots \qquad \cdot$$

$$Y_n = \beta_0 + \beta_1 X_{n1} + \beta_2 X_{n2} + \cdots + \beta_{K-1} X_{n, K-1}$$

From this we can produce the **X** matrix.

$$\mathbf{X} = \begin{bmatrix} 1 & X_{11} & X_{12} & \cdots & X_{1, K-1} \\ 1 & X_{21} & X_{22} & \cdots & X_{2, K-1} \\ \cdot & \cdot & \cdot & \cdots & \cdot \\ 1 & X_{n1} & X_{n2} & \cdots & X_{n, K-1} \end{bmatrix}$$

As an example let us take a bivariate situation with three observations of the explanatory variable. Let these observations be 1, 2, and 3. This would produce the following **X** matrix:

$$\mathbf{X} = \begin{bmatrix} 1 & 1 \\ 1 & 2 \\ 1 & 3 \end{bmatrix}$$

and the **X'X** matrix would be

$$\mathbf{X'X} = \begin{bmatrix} 3 & 6 \\ 6 & 14 \end{bmatrix}$$

The inverse of this matrix is

$$(\mathbf{X'X})^{-1} = \begin{bmatrix} \frac{14}{6} & -1 \\ -1 & \frac{1}{2} \end{bmatrix}$$

The reader should determine the inverse to see whether it corresponds to the result given here.

Finally, we can put forward some properties of inverted matrices.

(1) The inverse of $\mathbf{A}^{-1}$ is the matrix **A**. That is,

$$(\mathbf{A}^{-1})^{-1} = \mathbf{A}$$

(2) The inverse of a transpose is equal to the inverse transposed.

$$(\mathbf{A'})^{-1} = (\mathbf{A}^{-1})'$$

6.5 The rank of a matrix

In definitional terms the rank of a matrix is the number of vectors in the matrix which are linearly independent. For example, in the following matrix the second column is three times the first column.

$$A = \begin{bmatrix} 2 & 6 \\ 4 & 12 \\ 6 & 18 \end{bmatrix}$$

Since the second column is a linear function of the first column, the matrix has a rank of one.

Following on from our definition we can state that the rank of a matrix cannot exceed the number of rows or columns, whichever is smaller. In essence, when one of the rows or columns is linearly dependent upon another we can draw the conclusion that the matrix has redundant information. The row or column that is linearly dependent can be taken out of the matrix without losing any of the information in the matrix.

The quickest method to determine the rank of a matrix is to take its determinant. The order of the largest matrix with a nonzero determinant taken from any matrix gives us the rank of that matrix. Let us look at one or two examples. Take the matrix A from above. We know that determinants can only be obtained from square matrices and this matrix can produce three 2×2 submatrices. These are

$$A_1 = \begin{bmatrix} 2 & 6 \\ 4 & 12 \end{bmatrix} \quad A_2 = \begin{bmatrix} 4 & 12 \\ 6 & 18 \end{bmatrix} \quad A_3 = \begin{bmatrix} 2 & 6 \\ 6 & 18 \end{bmatrix}$$

To determine the rank of the matrix we find the determinant of each of these and if one of them is nonzero we know the rank of the matrix A is two. However, if all of them have a zero determinant we can conclude that the rank of matrix A is one. Only a matrix with all zeros has a rank of zero.

$$|A_1| = 0 \quad |A_2| = 0 \quad |A_3| = 0$$

All of the determinants are zero and we conclude that the rank of the matrix A is one. Let us take another example.

$$\mathbf{X} = \begin{bmatrix} 1 & 3 & 1 & 0 \\ 1 & 2 & 2 & 2 \\ 1 & 1 & 3 & 1 \end{bmatrix}$$

Since this is a 3 × 4 matrix, we cannot get a determinant for it. So we break it down into four 3 × 3 matrices. These are

$$\begin{bmatrix} 1 & 3 & 1 \\ 1 & 2 & 2 \\ 1 & 1 & 3 \end{bmatrix} \begin{bmatrix} 1 & 1 & 0 \\ 1 & 2 & 2 \\ 1 & 3 & 1 \end{bmatrix} \begin{bmatrix} 1 & 3 & 0 \\ 1 & 2 & 2 \\ 1 & 1 & 1 \end{bmatrix} \begin{bmatrix} 3 & 1 & 0 \\ 2 & 2 & 2 \\ 1 & 3 & 1 \end{bmatrix}$$

In seeking the determinant of each of these, we find that at least one of them is a nonzero determinant and can conclude that the rank of matrix **X** is three.

All of this is leading up to the following important property—important because of the dependence of regression upon the $(\mathbf{X'X})^{-1}$ matrix. If we multiply two matrices together where the rank of one equals a and the rank of the other equals b, the rank of the resulting matrix is not greater than a or b, whichever is smaller.

We can clarify this by using the preceding example. In this matrix, notice that the number of columns is greater than the number of rows; if this were a data matrix it would mean that we had as many explanatory variables as observations of each variable. Let us try to develop the $(\mathbf{X'X})^{-1}$ matrix.

$$\mathbf{X} = \begin{bmatrix} 1 & 3 & 1 & 0 \\ 1 & 2 & 2 & 2 \\ 1 & 1 & 3 & 1 \end{bmatrix} \quad \mathbf{X'} = \begin{bmatrix} 1 & 1 & 1 \\ 3 & 2 & 1 \\ 1 & 2 & 3 \\ 0 & 2 & 1 \end{bmatrix} \quad \mathbf{X'X} = \begin{bmatrix} 3 & 6 & 6 & 3 \\ 6 & 14 & 10 & 5 \\ 6 & 10 & 14 & 7 \\ 3 & 5 & 7 & 5 \end{bmatrix}$$

We wish to invert this matrix, and the first step in this process is to develop the determinant. But in attempting this we find that the determinant is zero, and thus we cannot obtain the inverse of the matrix $\mathbf{X'X}$. The rank of $\mathbf{X'X}$ is three, which is the same as the rank of **X**. In this data situation, we will not be able to obtain any estimates of the regression slopes.

We get into a similar situation when one of the explanatory variables is a linear function of another explanatory variable. The $\mathbf{X'X}$ matrix has a zero determinant and so we cannot invert it. This condition is known as multicollinearity, and we shall deal with the problem extensively at a later time.

Further readings

It would be helpful if the student worked his way through the examples in the following books on matrix algebra.

Goldberger: chapter 2.
Johnston: chapter 4.
Ayres, F. 1962. *Schaum's Outline of Theory and Problems of Matrices.* New York: McGraw-Hill.

7

multiple regression

We are ready to consider models in which there is more than one explanatory variable. In this situation, the model proposes that the explained variable is a function of a number of explanatory variables. This is a multiple regression model.

7.1 The multiple regression model

In the multiple regression model the basic regression equation is

$$Y_i = \beta_0 + \beta_1 X_{i1} + \beta_2 X_{i2} + \cdots + \beta_{K-1} X_{i, K-1} + u_i \tag{7.1}$$

In this model we have replaced α, the symbol for the regression intercept, with the symbol β_0. Each of the explanatory variables has two subscripts. The first refers to the observation number while the second refers to the variable number. There are a total of K variables in the equation—one explained variable and $K - 1$ explanatory variables. There are also K parameters in the model—one regression intercept and a regression slope for each of the explanatory variables.

We can develop the following set of equations, one for each observation or case in the regression:

$$
\begin{aligned}
\text{Case 1} \quad & Y_1 = \beta_0 + \beta_1 X_{11} + \beta_2 X_{12} + \cdots + \beta_{K-1} X_{1, K-1} + u_1 \\
\text{Case 2} \quad & Y_2 = \beta_0 + \beta_1 X_{21} + \beta_2 X_{22} + \cdots + \beta_{K-1} X_{2, K-1} + u_2 \\
& \cdot \qquad \cdot \qquad \cdot \qquad \cdot \qquad \cdots \qquad \qquad \cdot \\
\text{Case } n \quad & Y_n = \beta_0 + \beta_1 X_{n1} + \beta_2 X_{n2} + \cdots + \beta_{K-1} X_{n, K-1} + u_n
\end{aligned}
\tag{7.2}
$$

In terms of matrix notation, we see that the regression equation becomes

$$\mathbf{Y} = \mathbf{X}\boldsymbol{\beta} + \mathbf{u} \tag{7.3}$$

This is the general linear regression equation in matrix form where

$$\mathbf{Y} = \begin{bmatrix} Y_1 \\ Y_2 \\ \vdots \\ Y_n \end{bmatrix} \quad \mathbf{X} = \begin{bmatrix} 1 & X_{11} & X_{12} & \cdots & X_{1,K-1} \\ 1 & X_{21} & X_{22} & \cdots & X_{2,K-1} \\ \cdot & \cdot & \cdot & \cdots & \cdot \\ 1 & X_{n1} & X_{n2} & \cdots & X_{n,K-1} \end{bmatrix}$$

$$\boldsymbol{\beta} = \begin{bmatrix} \beta_0 \\ \beta_1 \\ \vdots \\ \beta_{K-1} \end{bmatrix} \quad \mathbf{u} = \begin{bmatrix} u_1 \\ u_2 \\ \vdots \\ u_n \end{bmatrix}$$

If we carry out the matrix multiplication suggested in equation (7.3) we will obtain the set of equations (7.2). We also see that the orders of the matrices above are as follows:

$\mathbf{Y}$ is an $n \times 1$ column vector

$\mathbf{X}$ is an $n \times K$ matrix

$\boldsymbol{\beta}$ is a $K \times 1$ column vector

$\mathbf{u}$ is an $n \times 1$ column vector

The column of ones in the $\mathbf{X}$ matrix allows the generation of the regression intercept.

7.2   The least-square assumptions in multiple regression

Having defined the regression model in matrix terms, let us consider the regression assumptions. As with the bivariate case, the criterion for locating the regression surface is the least-squares criterion. This requires the minimization of the sum of the error term squared, i.e., min $\Sigma\, u_i^2$. In matrix terms we wish to minimize $\mathbf{u'u}$ because

$$\mathbf{u} = \begin{bmatrix} u_1 \\ u_2 \\ \vdots \\ u_n \end{bmatrix} \quad \mathbf{u'} = \begin{bmatrix} u_1 & u_2 & \ldots & u_n \end{bmatrix}$$

$$\mathbf{u'u} = \begin{bmatrix} u_1^2 + u_2^2 + \cdots + u_n^2 \end{bmatrix} = \sum_{i=1}^{n} u_i^2$$

As with the bivariate case, in multiple regression the majority of the regression assumptions are concerned with the error term. The first five of the following seven assumptions are identical to the five we considered in Chapter 2. The sixth and seventh assumptions are peculiar to the multiple regression model.

Assumption 1. u is normally distributed.

Assumption 2. The mean of the error distribution is zero.

Assumption 3. The variance of the error distribution is constant over all values of the explanatory variables.

Assumption 4. The error terms for different values of the explanatory variables are independent of one another.

Assumption 5. All of the explanatory variables are nonstochastic and have a finite variance different from zero.

Assumption 6. The number of observations must exceed the number of coefficients to be estimated. That is, $n > K$.

Assumption 7. There must not be an exact linear relationship between any of the explanatory variables.

The final two assumptions are concerned with the rank of the $\mathbf{X'X}$ matrix and possibilities for the inversion required in calculations. After developing the calculating equations, we shall return to consider these two assumptions in more detail.

7.3   The normal equations in multiple regression

The development of the normal equations in the multiple regression model is quite complex. Instead of developing the equations, we shall simply state them and then show that for the bivariate case we obtain those normal equations presented in equations (2.6a) and (2.6b).

The least-square normal equations can be represented in matrix form as

$$(\mathbf{X'Y}) = (\mathbf{X'X})\hat{\boldsymbol{\beta}} \tag{7.4}$$

and when we premultiply by $(\mathbf{X'X})^{-1}$ we obtain

$$\hat{\boldsymbol{\beta}} = (\mathbf{X'X})^{-1}(\mathbf{X'Y}) \tag{7.5}$$

Let us now show that in the bivariate case this reduces to the normal equations presented earlier. This should encourage the reader into accepting the above relationship.

In the bivariate case we have the following matrices:

$$\mathbf{Y} = \begin{bmatrix} Y_1 \\ Y_2 \\ \vdots \\ Y_n \end{bmatrix} \quad \mathbf{X} = \begin{bmatrix} 1 & X_{11} \\ 1 & X_{21} \\ \vdots & \vdots \\ 1 & X_{n1} \end{bmatrix} \quad \mathbf{X}' = \begin{bmatrix} 1 & 1 & \cdots & 1 \\ X_{11} & X_{21} & \cdots & X_{n1} \end{bmatrix}$$

In developing the $(\mathbf{X}'\mathbf{X})$ matrix we get

$$(\mathbf{X}'\mathbf{X}) = \begin{bmatrix} n & \sum X_i \\ \sum X_i & \sum X_i^2 \end{bmatrix}$$

Also we see that $(\mathbf{X}'\mathbf{Y})$ is

$$\mathbf{X}'\mathbf{Y} = \begin{bmatrix} \sum Y_i \\ \sum X_i Y_i \end{bmatrix}$$

Putting this all together in equation (7.4), we finish with

$$\begin{bmatrix} \sum Y_i \\ \sum X_i Y_i \end{bmatrix} = \begin{bmatrix} n & \sum X_i \\ \sum X_i & \sum X_i^2 \end{bmatrix} \begin{bmatrix} \hat{\beta}_0 \\ \hat{\beta}_1 \end{bmatrix} \tag{7.6}$$

With multiplication this can be arranged to equal the normal equations developed in Chapter 2. Remember that $\hat{\beta}_0$ is the revised α.

$$\begin{bmatrix} \sum Y_i \\ \sum X_i Y_i \end{bmatrix} = \begin{bmatrix} \hat{\beta}_0 n + \hat{\beta}_1 \sum X_i \\ \hat{\beta}_0 \sum X_i + \hat{\beta}_1 \sum X_i^2 \end{bmatrix} \tag{7.7}$$

A similar technique would allow us to develop the normal equations for models with larger numbers of explanatory variables.

Let us try out one or two examples to become familiar with the actual calculating processes. First we use an example where there are two explanatory variables and five observations. Such a small sample size is selected so that students will be encouraged to try out the example for themselves.

Y	X_1	X_2
8	1	-2
5	-3	1
9	1	-2
5	0	0
-7	1	3

$$\mathbf{Y} = \begin{bmatrix} 8 \\ 5 \\ 9 \\ 5 \\ -7 \end{bmatrix} \quad \mathbf{X} = \begin{bmatrix} 1 & 1 & -2 \\ 1 & -3 & 1 \\ 1 & 1 & -2 \\ 1 & 0 & 0 \\ 1 & 1 & 3 \end{bmatrix}$$

$$(5 \times 3)$$

The X' and $X'X$ matrix are

$$X' = \begin{bmatrix} 1 & 1 & 1 & 1 & 1 \\ 1 & -3 & 1 & 0 & 1 \\ -2 & 1 & -2 & 0 & 3 \end{bmatrix} \quad X'X = \begin{bmatrix} 5 & 0 & 0 \\ 0 & 12 & -4 \\ 0 & -4 & 18 \end{bmatrix}$$

$$(3 \times 5) \qquad\qquad\qquad (3 \times 3)$$

The $X'Y$ matrix is

$$X'Y = \begin{bmatrix} 20 \\ -5 \\ -50 \end{bmatrix}$$

The inversion of $X'X$ produces

$$\begin{bmatrix} 0.2 & 0 & 0 \\ 0 & 0.09 & 0.02 \\ 0 & 0.02 & 0.06 \end{bmatrix}$$

$$(3 \times 3)$$

Using $\hat{\beta} = (X'X)^{-1}X'Y$, we obtain the estimates for β_0, β_1, and β_2.

$$\hat{\beta} = \begin{bmatrix} 0.2 & 0 & 0 \\ 0 & 0.09 & 0.02 \\ 0 & 0.02 & 0.06 \end{bmatrix} \begin{bmatrix} 20 \\ -5 \\ -50 \end{bmatrix} = \begin{bmatrix} 4 \\ -1.45 \\ -3.10 \end{bmatrix}$$

$$(3 \times 3) \qquad (3 \times 1) \qquad (3 \times 1)$$

So we see that the estimating line for these data is

$$\hat{Y}_i = 4.0 + (-1.45)X_{i1} + (-3.10)X_{i2}$$

Let us repeat the process for the percentage vote increase–money spent situation used in Chapter 2. The student will remember that the estimates in this bivariate situation as calculated at that time were $\beta_0 = -0.064$ and $\beta_1 = 0.002996$. The data were

Y_i	X_{i1}	Y_i	X_{i1}	Y_i	X_{i1}
2.8	1000	8.5	3000	14.7	5000
3.0	1000	9.0	3000	15.2	5000
3.1	1000	9.1	3000	15.5	5000
5.7	2000	11.6	4000	17.6	6000
6.0	2000	11.9	4000	18.0	6000
6.2	2000	11.9	4000	18.2	6000

Thus the relevant matrices are

$$
Y = \begin{bmatrix} 2.8 \\ 3.0 \\ 3.1 \\ 5.7 \\ 6.0 \\ 6.2 \\ 8.5 \\ 9.0 \\ 9.1 \\ 11.6 \\ 11.9 \\ 11.9 \\ 14.7 \\ 15.2 \\ 15.5 \\ 17.6 \\ 18.0 \\ 18.2 \end{bmatrix} \quad X = \begin{bmatrix} 1 & 1000 \\ 1 & 1000 \\ 1 & 1000 \\ 1 & 2000 \\ 1 & 2000 \\ 1 & 2000 \\ 1 & 3000 \\ 1 & 3000 \\ 1 & 3000 \\ 1 & 4000 \\ 1 & 4000 \\ 1 & 4000 \\ 1 & 5000 \\ 1 & 5000 \\ 1 & 5000 \\ 1 & 6000 \\ 1 & 6000 \\ 1 & 6000 \end{bmatrix} \quad X'X = \begin{bmatrix} 18 & 63 \times 10^3 \\ 63 \times 10^3 & 273 \times 10^6 \end{bmatrix}
$$

The $(X'X)^{-1}$ matrix is

$$
\begin{bmatrix} \dfrac{273}{945} & \dfrac{-63}{945 \times 10^3} \\ \dfrac{-63}{945 \times 10^3} & \dfrac{18}{945 \times 10^6} \end{bmatrix}
$$

The $X'Y$ matrix is

$$
X'Y = \begin{bmatrix} 187.6 \\ 813.9 \times 10^3 \end{bmatrix}
$$

On inserting all of this into $\hat{\beta} = (X'X)^{-1}X'Y$ we obtain our estimates

$$
\hat{\beta} = \begin{bmatrix} \dfrac{273}{945} & \dfrac{-63}{945 \times 10^3} \\ \dfrac{-63}{945 \times 10^3} & \dfrac{18}{945 \times 10^6} \end{bmatrix} \begin{bmatrix} 187.6 \\ 813.9 \times 10^3 \end{bmatrix} = \begin{bmatrix} -0.064 \\ 0.002996 \end{bmatrix}
$$

These are the same as those derived using the bivariate regular arithmetic method. This demonstrates that the matrix method for regression also obliges in situations where we have two variables.

7.4   The residuals in multiple regression

Again, we appear to have gotten ahead of ourselves, for as we have seen, the whole regression procedure revolves around the condition of the error distribution. As is usual, we do not know the distribution of the error term but have to rely upon the best estimates we have, the residuals from the regression procedure.

In matrix form the residuals e_i are given by the relationship

$$\mathbf{e} = \mathbf{Y} - \hat{\mathbf{Y}} \quad \text{where } \hat{\mathbf{Y}} \text{ is the } n \times 1 \text{ vector of estimated values}$$

But we also know that

$$\hat{\mathbf{Y}} = \mathbf{X}\hat{\boldsymbol{\beta}}$$

and so the residual matrix becomes

$$\mathbf{e} = \mathbf{Y} - \mathbf{X}\hat{\boldsymbol{\beta}} \tag{7.8}$$

Let us calculate the residuals for both of our examples. In the first example the relevant matrices are

$$\mathbf{Y} = \begin{bmatrix} 8 \\ 5 \\ 9 \\ 5 \\ -7 \end{bmatrix} \quad \mathbf{X} = \begin{bmatrix} 1 & 1 & -2 \\ 1 & -3 & 1 \\ 1 & 1 & -2 \\ 1 & 0 & 0 \\ 1 & 1 & 3 \end{bmatrix} \quad \hat{\boldsymbol{\beta}} = \begin{bmatrix} 4 \\ -1.45 \\ -3.10 \end{bmatrix}$$

When we insert these into matrix equation (7.8) we obtain

$$\mathbf{e} = \mathbf{Y} - \mathbf{X}\hat{\boldsymbol{\beta}}$$

$$\mathbf{e} = \begin{bmatrix} 8 \\ 5 \\ 9 \\ 5 \\ -7 \end{bmatrix} - \begin{bmatrix} 1 & 1 & -2 \\ 1 & -3 & 1 \\ 1 & 1 & -2 \\ 1 & 0 & 0 \\ 1 & 1 & 3 \end{bmatrix} \begin{bmatrix} 4.0 \\ -1.45 \\ -3.10 \end{bmatrix}$$

$$
= \begin{bmatrix} 8 \\ 5 \\ 9 \\ 5 \\ -7 \end{bmatrix} - \begin{bmatrix} 8.75 \\ 5.25 \\ 8.75 \\ 4.00 \\ -6.75 \end{bmatrix} = \begin{bmatrix} -0.75 \\ -0.25 \\ 0.25 \\ 1.00 \\ -0.25 \end{bmatrix}
$$

We can perform a similar operation on our second example. The relevant matrices are

$$
Y = \begin{bmatrix} 2.8 \\ 3.0 \\ 3.1 \\ 5.7 \\ 6.0 \\ 6.2 \\ 8.5 \\ 9.0 \\ 9.1 \\ 11.6 \\ 11.9 \\ 11.9 \\ 14.7 \\ 15.2 \\ 15.5 \\ 17.6 \\ 18.0 \\ 18.2 \end{bmatrix}
\quad
X = \begin{bmatrix} 1 & 1000 \\ 1 & 1000 \\ 1 & 1000 \\ 1 & 2000 \\ 1 & 2000 \\ 1 & 2000 \\ 1 & 3000 \\ 1 & 3000 \\ 1 & 3000 \\ 1 & 4000 \\ 1 & 4000 \\ 1 & 4000 \\ 1 & 5000 \\ 1 & 5000 \\ 1 & 5000 \\ 1 & 6000 \\ 1 & 6000 \\ 1 & 6000 \end{bmatrix}
\quad
\hat{\beta} = \begin{bmatrix} -0.064 \\ 0.002996 \end{bmatrix}
$$

and this produces the residual vector

$$
e = Y - X\hat{\beta} = \begin{bmatrix} -0.125 \\ 0.075 \\ 0.175 \\ -0.233 \\ 0.067 \\ 0.267 \\ -0.441 \end{bmatrix}
$$

(continued)

$$e = Y - X\hat{\beta} = \begin{vmatrix} 0.059 \\ 0.159 \\ -0.348 \\ -0.048 \\ -0.048 \\ -0.256 \\ 0.244 \\ 0.544 \\ -0.363 \\ 0.037 \\ 0.237 \end{vmatrix}$$

Having obtained the residual vector, we can use it to determine whether the classical assumptions have been violated, using the visual procedures described in Chapter 4. We can also use the residuals to generate the variance-covariance matrix of the regression coefficients.

7.5   The variance-covariance matrix of the regression coefficients

Let us take another look at the $X'X$ matrix for the bivariate situation. This matrix is

$$X'X = \begin{bmatrix} n & \sum X_i \\ \sum X_i & \sum X_i^2 \end{bmatrix}$$

If we take the inverse of this to produce $(X'X)^{-1}$ we obtain

$$(X'X)^{-1} = \begin{bmatrix} \dfrac{\sum X_i^2}{n \sum (X_i - \bar{X})^2} & \dfrac{-\bar{X}}{\sum (X_i - \bar{X})^2} \\ \dfrac{-\bar{X}}{\sum (X_i - \bar{X})^2} & \dfrac{1}{\sum (X_i - \bar{X})^2} \end{bmatrix}$$

From equations (3.3) and (3.4) we note that

$$\text{var}\,(\hat{\beta}_0) = \frac{\sigma^2 \sum X_i^2}{n \sum (X_i - \bar{X})^2}$$

and

$$\text{var}(\hat{\beta}_1) = \frac{\sigma^2}{\sum (X_i - \bar{X})^2}$$

If we compare these variances with the quantities in the top left-hand cell and the bottom right-hand cell of the $(\mathbf{X'X})^{-1}$ matrix, we see that the only difference is σ^2, the variance of the error term. In this roundabout manner we have shown that the variance-covariance matrix of the regression coefficients is given by

$$\text{variance-covariance of regression coefficients} = \sigma^2(\mathbf{X'X})^{-1} \quad (7.9)$$

The quantities in the top left to bottom right diagonal are the variance of the regression coefficients, while the off-diagonal quantities are the covariances. The matrix shows this. Notice that it's a $K \times K$ matrix.

$$\begin{matrix} \text{variance-} \\ \text{covariance} \\ \text{of regression} \\ \text{coefficients} \end{matrix} = \begin{bmatrix} \text{var}(\beta_0) & \text{cov}(\beta_0, \beta_1) & \text{cov}(\beta_0, \beta_2) & \cdots & \text{cov}(\beta_0, \beta_{K-1}) \\ \text{cov}(\beta_0, \beta_1) & \text{var}(\beta_1) & \text{cov}(\beta_1, \beta_2) & \cdots & \text{cov}(\beta_1, \beta_{K-1}) \\ \text{cov}(\beta_0, \beta_2) & \text{cov}(\beta_1, \beta_2) & \text{var}(\beta_2) & \cdots & \text{cov}(\beta_2, \beta_{K-1}) \\ \vdots & \vdots & \vdots & & \vdots \\ \text{cov}(\beta_0, \beta_{K-1}) & \cdots & & \cdots & \cdots & \text{var}(\beta_{K-1}) \end{bmatrix}$$

$$(K \times K)$$

Of course, we do not have σ^2 and must estimate it using the usual equation

$$\hat{\sigma}^2 = \frac{\sum e_i^2}{n - K}$$

and if we restructure this in matrix format we get

$$\hat{\sigma}^2 = \frac{\mathbf{e'e}}{n - K} \quad (7.10)$$

We arrive at this because

$$\mathbf{e} = \begin{bmatrix} e_1 \\ e_2 \\ \vdots \\ e_n \end{bmatrix} \quad \text{and} \quad \mathbf{e'} = [e_1 \quad e_2 \quad e_3 \quad \cdots \quad e_n]$$

Thus

$$\mathbf{e'e} = e_1{}^2 + e_2{}^2 + e_3{}^2 + \cdots + e_n{}^2 = \sum e_i{}^2$$

The final relationship for obtaining the estimates of the variance-covariance matrix for the regression coefficients is

$$\hat{\sigma}^2_{(\beta)} = \frac{\mathbf{e'e}}{n - K} (\mathbf{X'X})^{-1} \tag{7.11}$$

We can now calculate the variance-covariance matrix for the first of our examples. In this first example we have the following constituents:

$$(\mathbf{X'X})^{-1} = \begin{bmatrix} 0.2 & 0 & 0 \\ 0 & 0.09 & 0.02 \\ 0 & 0.02 & 0.06 \end{bmatrix} \quad \mathbf{e} = \begin{bmatrix} -0.75 \\ -0.25 \\ 0.25 \\ 1.00 \\ -0.25 \end{bmatrix}$$

from this we can obtain $\mathbf{e'e}$:

$$\mathbf{e'e} = \begin{bmatrix} -0.75 & -0.25 & 0.25 & 1.00 & -0.25 \end{bmatrix} \begin{bmatrix} -0.75 \\ -0.25 \\ 0.25 \\ 1.00 \\ -0.25 \end{bmatrix} = 1.75$$

Putting this into equation (7.11),

$$\hat{\sigma}^2_{(\beta)} = \frac{\mathbf{e'e}}{n - K} (\mathbf{X'X})^{-1} = \frac{1.75}{5 - 3} \begin{bmatrix} 0.2 & 0 & 0 \\ 0 & 0.09 & 0.02 \\ 0 & 0.02 & 0.06 \end{bmatrix}$$

$$= \begin{bmatrix} 0.175 & 0 & 0 \\ 0 & 0.078 & 0.0175 \\ 0 & 0.0175 & 0.0525 \end{bmatrix}$$

From this we see that the variance of $\hat{\beta}_0 = 0.175$, the variance of $\hat{\beta}_1 = 0.078$, and the variance of $\hat{\beta}_2 = 0.0525$. We need these quantities to test hypotheses about the regression estimates.

7.6   The coefficient of determination

In discussing regression in the first part of the book, we dealt with the correlation coefficient as a measure of goodness of fit for the regression line. There is just one fundamental difficulty with the simple correlation coefficient as a measure of the "goodness of fit" of any line—it can only accommodate situations in which there is one explanatory variable. Thus when we move into situations where there is more than one explanatory variable, this measure is rendered impotent. We can, however, develop a measure which is similar to the simple correlation coefficient, in that it measures the correlation between both sides of the regression equation, whatever the number of explanatory variables. This is called the coefficient of determination (or multiple correlation or R^2). And it is generous enough to reduce to the square of the simple correlation when there is only one explanatory variable.

The coefficient of determination can be developed by decomposing the variance of the explained variable Y_i, and we shall carry out this development for the bivariate case, as a teaching ploy and because the same argument can be made for the multivariate case.

Consider diagram 7.1. The total variation of Y_i can be represented by the difference between each observation of Y_i and the mean of the Y_i distribution, $\bar{Y}$. The variance of Y_i is defined by the term $\Sigma (Y_i - \bar{Y})^2/n$. Now suppose that, with the trickery learned in Chapter 2, we locate a least-squares regression line through these data points. The variation of Y_i is a function of two components. The first is due to the relationship between Y and X. This is the regression function. And the second is

Diagram 7.1

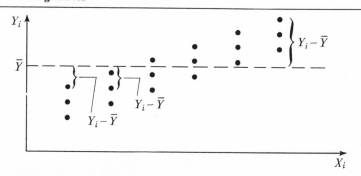

due to the other excluded variables acting on Y_i. This is the error function. Thus we have some of the variance of Y explained by the regression line and some unexplained by the regression line. Diagram 7.2 shows both of these components of the error. Together they make up the total variance of Y_i. Thus we can put down the following verbal equation, for the variance of the Y observations,

$$\text{total variance of } Y_i = \frac{\text{variance explained}}{\text{by regression}} + \frac{\text{variance unexplained}}{\text{by regression}}$$

or more rigorously,

$$\begin{array}{ccc} \text{total sum} & \text{regression sum} & \text{error sum} \\ \text{of squares} = & \text{of squares} & + \text{of squares} \\ \text{(SST)} & \text{(SSR)} & \text{(SSE)} \end{array} \qquad (7.12)$$

and mathematically,

$$\sum (Y_i - \bar{Y})^2 = \sum (\hat{Y}_i - \bar{Y})^2 + \sum (Y_i - \hat{Y}_i)^2 \qquad (7.13)$$

Although we have developed this in a quasi-mathematical fashion, it can be shown to be mathematically correct.

Diagram 7.2

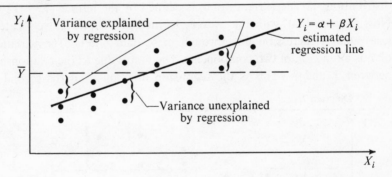

Returning to the main reason for all of this, we are looking for some measure of the goodness of fit of the regression line to the data. Obviously, since we are interested in explaining Y_i, the more of the variance of Y_i explained by the regression line the better we like our regression line. We would have a perfect regression line if all of the variance of Y_i was explained by the regression line. So the ratio of variance of Y_i explained by the regression line to the total variance of Y_i is some measure of the goodness of our regression line. We define the coefficient of determination.

$$\text{coefficient of determination } R^2 = \frac{\text{SSR}}{\text{SST}} \qquad (7.14)$$

By rearranging equation (7.12) we get

$$\frac{\text{SST}}{\text{SST}} = \frac{\text{SSR}}{\text{SST}} + \frac{\text{SSE}}{\text{SST}} \qquad (7.15)$$

thus

$$R^2 = 1 - \frac{\text{SSE}}{\text{SST}} \qquad (7.16)$$

Looking at this equation, we see that the maximum value of $R^2 = 1$. When there are no errors from the regression line, all of the variance of Y_i is explained by the regression line. The lowest value of R^2 is zero, when the regression line explains none of the variance of Y_i. Unlike the simple correlation coefficient, it is not capable of attaining negative quantities. Let us calculate the coefficient of determination for our money-spent–percentage-vote example.

Combining equations (7.13) and (7.16),

$$R^2 = 1 - \frac{\sum (Y_i - \hat{Y})^2}{\sum (Y_i - \bar{Y})^2} \qquad (7.17)$$

we see a need to calculate the estimated values of Y_i. This is done by using the estimated equation for the example,

$$\hat{Y}_i = -0.064 + \frac{2.9962}{1000} (X_i)$$

Table 7.1 contains all of the data for the R^2 calculation. Notice that the quantity is 0.9982, the same as the simple correlation coefficient squared. The reader should fully understand that this only occurs when there is one explanatory variable.

Having developed this measure of "goodness of fit," we should discuss it. If we obtain a low value for R^2, it tells us that our model is quite limited. It means that little of the variation of the explained variable Y_i is accounted for by the explanatory variables. Alternatively, it can suggest that the

Table 7.1

Y_i	$\hat{Y}_i$	$e_i = Y_i - \hat{Y}_i$	$(Y_i - \hat{Y}_i)^2$	$(Y_i - \bar{Y})^2$
2.80	2.93	−0.13	.0169	58.06
3.00	2.93	0.07	.0049	55.06
3.10	2.93	0.17	.0289	53.58
5.70	5.93	−0.23	.0529	22.28
6.00	5.93	0.07	.0049	19.54
6.20	5.93	0.27	.0729	17.81
8.50	8.92	−0.43	.1849	3.69
9.00	8.92	0.07	.0049	2.02
9.10	8.92	0.17	.0289	1.74
11.60	11.92	−0.33	.1089	1.39
11.90	11.92	−0.03	.0009	2.19
11.90	11.92	−0.03	.0009	2.19
14.70	14.92	−0.23	.0529	18.32
15.20	14.92	0.27	.0729	22.85
15.10	14.92	0.17	.0289	21.90
17.60	17.91	−0.33	.1089	51.55
18.00	17.91	0.07	.0049	57.46
18.20	17.91	0.27	.0729	60.53
			.8513	472.16

$$\bar{Y} = \frac{187.6}{18} = 10.42 \qquad R^2 = 1 - \frac{.8513}{472.16} = 0.9982$$

error component in our equation (which, if you remember, reflects those causes of Y_i that we have left out of the model) is large.

Another interpretation of R^2 is that it is a direct measure of the percentage explanation in the model. In our example we have accounted for 99.82 percent of the variance of the explained variable. Possible explanations of low R^2 are therefore (1) imprecise measurement of the explained variable and (2) misspecified regression model, both of which are imperfections in our model and which we rather hope the measure of goodness of fit will detect.

The advantage of the R^2 value is that it can cope with situations where there is more than one explanatory variable. The simple correlation coefficient cannot do this. Secondly the R^2 value gives us an absolute amount of variance accounted for by the model. This value ranges from 0 (0 percent) to 1 (100 percent). Thirdly, it is very easy to develop tests of significance for the R^2 value. This, as we shall see in the next chapter, is important because it allows us to test the significance of a whole model at once instead of each regression slope individually.

The coefficient of determination has one disadvantage. It can be used by unscrupulous empiricists as a device for fishing out variables to add to any regression equation to increase the R^2 value. It is to stop this inductive epidemic that the corrected coefficient of determination was developed.

7.7   The corrected coefficient of determination

It is usually the case that the addition of variables to any regression equation will increase the R^2 value. In any case, no additional variable can reduce the R^2 value. Thus, it has been used by researchers to "fish" for variables to add to an already existing equation. We get into the situation where the researcher is adding variables inductively rather than by *a priori* theoretical reasoning. To be sure, these empiricists can usually find some *post hoc* justification for adding the variables.

This difficulty can be slightly guarded against by providing a corrected coefficient of determination which takes into account not only the explained variance but also the number of variables in the equation and the number of observations from which the estimates are procured.

The aim of the corrected coefficient is to prevent the comparison of models purely on the amount of variation explained, when the models being compared have differing numbers of explanatory variables and are confronted with different sample sizes. Using our corrected equation it is possible to *reduce* the corrected R^2 by adding variables to an equation. The equation for the corrected coefficient of determination $\bar{R}^2$ is

$$\bar{R}^2 = R^2 - \frac{K - 1}{n - K}(1 - R^2) \tag{7.18}$$

where K is the number of variables both explained and explanatory in the regression model and n is the sample size.

It should be noted that the term $K - 1$ is the number of explanatory variables in the model and that the term $n - K$ is the number of degrees of freedom. The equation is arranged such that

$$\bar{R}^2 \leqslant R^2$$

Using our example when $n = 18$ and $K = 2$, our $\bar{R}^2$ is given by

$$\bar{R}^2 = 0.9982 - \frac{2 - 1}{18 - 2}(1 - 0.9982) = 0.9981$$

What this corrected coefficient does is charge us for every explanatory variable we add for a given number of observations. Mostly we are in situations where our number of observations is fixed. However, the number of variables in the regression need not be. We should pay some cost for the addition of every explanatory variable. Obviously it costs us in parsimony, but more than this, with the corrected coefficient it also costs us in the amount of variation accounted for. This, in the final analysis, will reflect upon the appropriateness of our model.

7.8　The coefficient of determination: matrix formulation

It can be shown that equation (7.13) expressed in matrix terms becomes

$$\mathbf{Y'Y} - \frac{(\sum Y_i)^2}{n} = \boldsymbol{\beta}'\mathbf{X'Y} - \frac{(\sum Y_i)^2}{n} + \mathbf{e'e} \qquad (7.19)$$

Notice that there are two terms exactly the same on each side of the equation, and one might be tempted to cancel them out. This should not be done because when we develop R^2, the coefficient of determination, which is given by $1 - \text{SSE/SST}$, we would obtain erroneous values. We see that in matrix formulation the coefficient of determination is given by the relationship

$$R^2 = 1 - \frac{\mathbf{e'e}}{\mathbf{Y'Y} - \dfrac{(\sum Y_i)^2}{n}} \qquad (7.20)$$

Let us calculate the coefficient of determination in our small sample example. The only terms we require here that we have not calculated already are $\mathbf{Y'Y}$ and $(\sum Y)^2/n$; $\mathbf{e'e}$ was calculated in Section 7.4.

$$\mathbf{Y'} = \begin{bmatrix} 8 & 5 & 9 & 5 & -7 \end{bmatrix} \quad \mathbf{Y} = \begin{bmatrix} 8 \\ 5 \\ 9 \\ 5 \\ -7 \end{bmatrix}$$

$$\mathbf{Y'Y} = 64 + 25 + 81 + 25 + 49 = 244$$

$$\frac{(\sum Y)^2}{n} = \frac{(20)^2}{5} = 80$$

Thus we get for our coefficient of determination

$$R^2 = 1 - \frac{1.75}{244 - 80} = 1 - 0.0106$$
$$= 0.9893$$

The calculation of the R^2 value, using matrix algebra for our money-spent–percentage-vote example is left as an exercise for the student.

Notice that the R^2 equation, although couched in matrix terms, does not affect the establishment of the corrected coefficient of determination $\bar{R}^2$, which is derived from the regular coefficient of determination.

7.9 A discussion of the added classical assumptions

In maturing from bivariate to multivariate regression, we added two assumptions to the existing five. These two were

(1) The number of observations must exceed the number of coefficients to be estimated. That is, $n > K$.
(2) There must not be any exact linear relationship between the explanatory variables.

Let us consider both of these.

7.9.1 Degrees of freedom problem

One can provide two arguments for this requirement. If the number of observations is less than or equal to the number of coefficients to be determined, the following is true:

$$n \leqslant K \quad \text{and so} \quad n - K \leqslant 0 \quad \text{i.e., the number of degrees of freedom is zero or less}$$

In this situation, the estimated variance of the error term, which is given by the relationship $\mathbf{e}'\mathbf{e}/(n - K)$, becomes infinity if $n = K$ and negative if $n < K$. Neither situation is acceptable.

The second argument concerns the rank of the $\mathbf{X}'\mathbf{X}$ matrix. In Chapter 6, in dealing with the rank of a matrix, we declared the following two rules:

(1) In any $m \times n$ matrix, the rank of that matrix is less than or equal to m or n, whichever is the smaller. So if we have an $\mathbf{X}$ matrix of $n \times K$ where n is less than K, the matrix will have at best a rank of n.

(2) If we multiply two matrices of rank a and b, the rank of the resulting matrix will be less than or equal to a or b, whichever is smaller. Thus if we develop the $\mathbf{X'X}$ matrix when $n < K$, it will have a rank of n at most.

But the $\mathbf{X'X}$ matrix will have an order of $K \times K$, and when we calculate the determinant of this matrix we will get $|\mathbf{X'X}| = 0$ because the rank of the matrix is less than the order of the matrix. Thus we cannot establish an inverse of $\mathbf{X'X}$. This is a fatal blow to our estimating procedure since the $(\mathbf{X'X})^{-1}$ is fundamental to the calculating process.

We see that when n is less than or equal to K we are unable to proceed in establishing the various quantities. Let us exploit a fictitious example where $n = 3$ and $K = 4$. The $\mathbf{X}$ matrix is

$$\mathbf{X} = \begin{bmatrix} 1 & 1 & 3 & 6 \\ 1 & 2 & 1 & 2 \\ 1 & 3 & 3 & 1 \end{bmatrix}$$

This matrix has, at most, a rank of 3. The $\mathbf{X'X}$ matrix is given by

$$\mathbf{X'X} = \begin{bmatrix} 3 & 6 & 7 & 9 \\ 6 & 14 & 14 & 13 \\ 7 & 14 & 19 & 23 \\ 9 & 13 & 23 & 41 \end{bmatrix}$$

This matrix can only have a rank of 3. This being so, the determinant of $\mathbf{X'X}$ will be zero. The student should check that this is true. The condition renders futile any further steps in the inversion of the $\mathbf{X'X}$ matrix.

7.9.2   Multicollinearity

The argument in Section 7.9.1 requires that the $\mathbf{X}$ matrix must have a rank of K. If it does not, then $\mathbf{X'X}$ will have a rank of less than K, with the result that the determinant of $\mathbf{X'X}$ will be zero and it becomes impossible to invert $\mathbf{X'X}$. This condition also exists when any one column in the $\mathbf{X}$ matrix is linearly dependent upon another. When at least one column is linearly dependent upon another, we will not be able to invert the consequent $\mathbf{X'X}$ matrix and thus will be unable to develop any estimates. This condition is called collinearity and exists when any one variable is perfectly correlated with any one other. When one variable is perfectly correlated with a linear set of the others, the condition is called *multicollinearity*.

Let us take another fictitious example to demonstrate. In the following matrix the second column is twice the third column, indicating a perfect linear relationship between the two.

$$\mathbf{X} = \begin{bmatrix} 1 & 2 & 1 \\ 1 & 4 & 2 \\ 1 & 6 & 3 \\ 1 & 8 & 4 \end{bmatrix}$$

In developing the $\mathbf{X'X}$ matrix we obtain

$$\mathbf{X'X} = \begin{bmatrix} 4 & 20 & 10 \\ 20 & 120 & 60 \\ 10 & 60 & 30 \end{bmatrix}$$

The determinant of this is zero. Inversion of the $\mathbf{X'X}$ matrix is thus impossible.

We can describe two extreme positions of multicollinearity. There is the condition above, called perfect multicollinearity, in which nothing can be done. At the other extreme we have the position of no multicollinearity. There are no problems in this situation, but unfortunately in nonexperimental sciences it does not occur. Mostly we end up with *some* multicollinearity, and so it becomes a question of degree.

Substantively, multicollinearity means that we cannot separate out the individual effects of the explanatory variables which are related. Let us take the classic example of the end-of-work whistle in London and Edinburgh. If both whistles go off at one o'clock and the workers leave both factories for lunch, we cannot tell which whistle caused which workers to leave which factory. Only by restraining one of the whistles and observing the effect on the workers' actions can we break this statistical deadlock. As this example suggests, when we have perfect collinearity or multicollinearity we require some extra information to break the deadlock.

Since we do not always get perfect multicollinearity, we should try to determine the effects on our parameters and their variances for different degrees of the disease. The most dramatic effect is upon the variance of the regression estimates, which explode as the multicollinearity approaches perfection.

Johnston[1] has shown the following explosion of the variances of the regression estimates.

	α				
	0	0.5	0.9	0.99	1.00
Variance of regression coefficients	$\sigma^2_{(\beta)}$	$\frac{4}{3}\sigma^2_{(\beta)}$	$5\sigma^2_{(\beta)}$	$\doteq 100\sigma^2_{(\beta)}$	∞

In that chart, α is the correlation between the two explanatory variables in the following multiple regression equation:

$$Y_i = \beta_0 + \beta_1 X_{i1} + \beta_2 X_{i2} + u_i$$

The student will perceive the tremendous explosion of the variance as the collinearity increases. The effect of this is felt particularly acutely in our testing procedure. Multicollinearity has no effect on the unbiased quality of our estimate but severely reduces the efficiency of the estimator. The student will remember that, as the variance of the sampling distribution increases, so does our likelihood of accepting the null-hypothesis. We may thus make incorrect inferences. And although we may have an unbiased estimate of the regression coefficients, it isn't much comfort because the large variance will increases the probability that our estimate is not close to the target population parameter.

Obvious questions concern the detection and remedy for multicollinearity. The first is relatively easy while the second is almost impossible. An overall criterion has been developed for detecting multicollinearity, and it is worth using this as a rule of thumb in situations where multicollinearity is suspected. Unfortunately, the student is not quite ready to discuss means of detection until after contemplating hypothesis-testing in multivariate situations. This is done in the next chapter, and we shall discuss multicollinearity detection at that time.

While exposure of the problem, as we shall see, is relatively easy, its solution is not. The most usual maneuver is to drop the explanatory variable causing the problem out of the regression equation. But as we shall see, it introduces bias into the other estimates if a variable which should rightfully be there is left out of the regression equation. The choice

1. J. Johnston, *Econometric Methods*, p. 161.

thus is between bias in the estimates (when we leave a variable out) or inefficiency (when we leave a variable in). The researcher must decide which is the lesser of two evils here, and the decision will depend upon the use of the research.

In summary, there are three multicollinearity situations. In the first, there is no multicollinearity and all the explanatory variables are independent of one another. This is the ideal situation. There is perfect multicollinearity. In this case no $(X'X)^{-1}$ can be produced, and we will not be able to proceed. However, the most usual case is somewhere in between these two extremes. The degree of multicollinearity allowable depends upon the research situation.

Mostly, nothing can be done about multicollinearity. The culprit(s) may be left out of the equation but this introduces bias into the estimates of the regression slopes remaining. If the culprit(s) are left in, we become vulnerable to type II errors. The only substantial correction is to gather more information about the explanatory variables. Since the major difficulty is disentangling the various effects, information that can help in this is required.

Further readings

Substantive
Black, E., and M. Black. 1973. "The Wallace Vote in Alabama: A Multiple Regression Analysis." *Journal of Politics* 35; 730.
Forbes, H. D., and E. R. Tufte. 1968. "A Note of Caution in Causal Modelling." *Am. Pol. Sci. Review* 62 : 1258.
Iverson, G. R. 1972. "Social Sciences and Statistics." *World Politics* 25: 145.
Kramer, G. H. 1971. "Short-Term Fluctuations in U.S. Voting Behavior, 1896–1964." *Am. Pol. Sci. Review* 65: 131.
Paranzino, D. 1972. "Inequality and Insurgency in Vietnam: A Further Re-Analysis." *World Politics* 24: 565.

Statistical
Goldberger: chapter 4.
Johnston: chapter 5.
Kmenta: chapter 10.
Wonnacott and Wonnacott: pp. 237–48.

8

various testing sequences: multivariate situations

In Chapter 3 we presented ways of testing hypotheses about individual regression coefficients, both intercept and slope. We will again do all of this but will present the argument employing a multivariate perspective. Added to this we will present techniques for testing the adequacy of whole lines rather than individual regression slopes.

8.1 Testing various hypotheses about individual regression slopes

We can carry out any or all of the following hypothesis-testing procedures:

(1) test whether any particular regression estimate is different from a prescribed value,
(2) test whether any one regression slope in a regression equation is different from another regression slope in the same equation,
(3) test whether two regression slopes from the same equation add together to produce some specific constant quantity.

We will take each of these in order and illustrate with examples.

8.1.1 Basic regression estimate testing

Any hypothesis-testing process, confidence interval, or prediction calculation requires the prior determination of the variance-covariance matrix of the estimator concerned. In our case the variance-covariance matrix of $\hat{\beta}$ is required, and as we saw in the previous chapter it is given by

$$\text{var-cov}(\beta) = \sigma^2(\mathbf{X'X})^{-1}$$

And when we insert our best estimate of σ^2 we obtain

$$\text{var-cov}(\hat{\beta}) = \frac{\mathbf{e'e}}{n-K}(\mathbf{X'X})^{-1}$$

When we lay out this matrix fully we produce

$$\text{var-cov}(\beta) = \begin{bmatrix} \text{var}(\beta_0) & \text{cov}(\beta_0, \beta_1) & \cdots & \text{cov}(\beta_0, \beta_{K-1}) \\ \text{cov}(\beta_1, \beta_0) & \cdots & & \cdots \\ \cdot & \cdots & \cdots & \cdot \\ \cdot & \cdots & \cdots & \cdot \\ \cdot & & \cdots & \cdot \\ \text{cov}(\beta_{K-1}, \beta_0) & \cdots & \cdots & \text{var}(\beta_{K-1}) \end{bmatrix}$$

The variances shown in the top left to bottom right diagonal are those required for the test statistic

$$\frac{\hat{\beta}_k - \text{constant}}{\sqrt{\text{var}(\hat{\beta}_k)}} \stackrel{\text{d}}{=} t \quad \text{with } n - K \text{ degrees of freedom}$$

where $\hat{\beta}_k$ = the regression estimate under examination
constant = value with which we wish to compare $\hat{\beta}_k$
$\sqrt{\text{var}(\hat{\beta}_k)} = \text{SD}(\hat{\beta}_k)$

The right-hand side of the equation can be interpreted as part of the whole equation, that is, the statistic developed on the left-hand side is distributed as a t-distribution with $n - K$ degrees of freedom.

For example, suppose that we have a regression equation

$$Y_i = \beta_0 + \beta_1 X_{i1} + \beta_2 X_{i2} + \beta_3 X_{i3} + u_i$$

and we wish to test if the regression slope associated with X_2 is different from zero. The hypothesis would be

$$H_0 : \hat{\beta}_2 = \beta_2 = 0$$

and the test statistic is

$$t = \frac{\hat{\beta}_2 - \beta_2}{\sqrt{\text{var}(\hat{\beta}_2)}} \quad \text{with } n - 4 \text{ degrees of freedom}$$

Let us insert some figures here. Let us say that $\hat{\beta}_2 = 0.35$ and that the variance of $\hat{\beta}_2$ taken from the variance-covariance matrix is 0.04. The sample size is 25. The test statistic becomes

$$t = \frac{0.35 - 0.0}{\sqrt{0.04}} = 1.75 \quad \text{with 21 degrees of freedom}$$

This is significant at the 0.05 level, one-tailed, allowing us to reject the null-hypothesis and accept the alternative.

As a second example let us return to the data given in the previous chapter. In that example we had the following data:

Y	X_1	X_2
8	1	−2
5	−3	1
9	1	−2
5	0	0
−7	1	3

Upon regression this produced the following estimated line

$$\hat{Y} = 4 - 1.45X_1 - 3.10X_2$$

The variance-covariance matrix is

$$\sigma^2_{(\hat{\beta})} = \begin{bmatrix} 0.175 & 0 & 0 \\ 0 & 0.078 & 0.0175 \\ 0 & 0.0175 & 0.0525 \end{bmatrix}$$

Using this, let us test various hypotheses concerning each of the regression estimates.

(1) Does $\hat{\beta}_0$ differ from zero, i.e., does the line go through the origin?

$$H_0: \hat{\beta}_0 = 0$$

If this is rejected, we can accept that the regression line does not go through the origin. The test statistic is

$$t = \frac{4.0 - 0.0}{\sqrt{0.175}} = 9.57 \quad \text{with 2 degrees of freedom}$$

Thus we can reject the null-hypothesis at the 0.05 one-tailed significance level.

(2) Does $\hat{\beta}_1$ differ from zero, i.e., does the variable X_1 influence the explained variable Y?

$$H_0: \hat{\beta}_1 = 0$$

If we reject this hypothesis, we can accept the alternative inference that X_1 does influence the explained variable. The test statistic is

$$t = \frac{-1.45 - 0.0}{\sqrt{0.078}} = -5.197 \quad \text{with 2 degrees of freedom}$$

We reject the null-hypothesis at the 0.05 one-tailed significance level.

(3) Does $\hat{\beta}_2$ differ from a value of -3.50, a value we may have arrived at theoretically?

$$H_0: \hat{\beta}_2 = -3.50$$

The test statistic for this is

$$t = \frac{-3.10 - (-3.50)}{\sqrt{0.0525}} = 1.433 \quad \text{with 2 degrees of freedom}$$

We accept the null-hypothesis, implying that the estimated slope is close to the one we would have predicted theoretically.

Reference to Chapter 3 shows that we have added little to the testing process described there. The fundamental difference is that we have plucked the variances for the test statistic from the variance-covariance matrix rather than from equations given at that time.

8.1.2   Hypotheses about regression slopes from the same equation

There are some research situations where the researcher needs to test whether there are differences between regression slopes from the same regression equation. Such hypotheses might be determined for theoretical considerations. In a general form this would be stated as

$$H_0: \hat{\beta}_j = \hat{\beta}_k \quad \text{or} \quad \hat{\beta}_j - \hat{\beta}_k = 0$$

The general test statistic for this would be

$$t = \frac{\beta_j - \beta_k}{\sqrt{\text{var}(\beta_j) + \text{var}(\beta_k) - 2\,\text{cov}(\beta_j, \beta_k)}} \overset{\text{d}}{=} t \quad \begin{array}{l} \text{with } n - K \\ \text{degrees of} \\ \text{freedom} \end{array} \quad (8.1)$$

While the top line of this statistic will be familiar to the student, the bottom may not be. The best way to describe how this comes about is to consider the variance of $\hat{\beta}_j$ as a^2 and the variance of $\hat{\beta}_k$ as b^2. In reality when we want the combined variance of $\hat{\beta}_j$ and $\hat{\beta}_k$ we want $(a + b)^2$. This, when expanded, becomes $a^2 + b^2 + 2ab$. Returning to variance language this becomes variance of $\hat{\beta}_j$ plus the variance of $\hat{\beta}_k$ plus twice the covariance of $\hat{\beta}_j$ and $\hat{\beta}_k$. In our particular hypothesis, however, we want the variance of $\hat{\beta}_j - \hat{\beta}_k$. Going back to our transformation, this is equal to $(a - b)^2$, which on multiplication becomes $a^2 + b^2 - 2ab$. Re-transformation produces variance of $\hat{\beta}_j$, plus the variance of $\hat{\beta}_k$ minus the covariance of $\hat{\beta}_j$ and $\hat{\beta}_k$, our denominator in the equation for the test statistic.

Let us use our example to illustrate. Our hypothesis is that the regression slope $\hat{\beta}_2$ is not different from the regression slope $\hat{\beta}_1$. This can be stated formally as

$$H_0 : \hat{\beta}_1 - \hat{\beta}_2 = 0$$

The test statistic is

$$t = \frac{\hat{\beta}_1 - \hat{\beta}_2}{\sqrt{\text{var}\,(\hat{\beta}_1) + \text{var}\,(\hat{\beta}_2) - 2\,\text{cov}\,(\hat{\beta}_1, \hat{\beta}_2)}}$$

When we supply the numerical values in the appropriate places, this becomes

$$t = \frac{-1.45 - (-3.10)}{\sqrt{0.078 + 0.0525 - 2(0.0175)}} = 5.35$$

This is significant at the 0.05 level, and we can conclude that there is a significant statistical difference between the two regression slopes. For testing whether more than two regression slopes are equal, the student is referred to Goldberger.[1]

8.1.3 Hypotheses concerning the aggregation of two regression slopes

Such hypotheses are related to that considered in the preceding section. In essence, the hypothesis is of the general form

$$H_0 : \hat{\beta}_j + \hat{\beta}_k = \text{constant}$$

1. A. S. Goldberger, *Econometric Theory*, pp. 172 ff.

The value of the constant can be anything, including zero, and although addition is shown here, subtraction of regression slopes to equal some constant is also allowed. This was shown in the Section 8.1.2 and we see that the hypothesis tested there is a special case of this more general hypothesis. The general test statistic is

$$t = \frac{\hat{\beta}_j + \hat{\beta}_k - \text{constant}}{\sqrt{\text{var}(\hat{\beta}_j) + \text{var}(\hat{\beta}_k) + 2\,\text{cov}(\hat{\beta}_j, \hat{\beta}_k)}} \overset{d}{=} t \quad \begin{array}{l} \text{with } n - K \\ \text{degrees of} \\ \text{freedom} \end{array} \quad (8.2)$$

Let us test the following hypothesis as an example:

$$H_0 = \hat{\beta}_1 + \hat{\beta}_2 = -4.0$$

We can imagine that we have arrived at the -4.0 value from some theoretical position. The test statistic is

$$t = \frac{\hat{\beta}_1 + \hat{\beta}_2 - (-4.0)}{\sqrt{\text{var}(\hat{\beta}_1) + \text{var}(\hat{\beta}_2) + 2\,\text{cov}(\hat{\beta}_1, \hat{\beta}_2)}}$$

Plucking the various values from the variance-covariance matrix we get

$$t = \frac{-1.45 + (-3.10) - (-4)}{\sqrt{0.078 + 0.0525 + 2(0.0175)}} = -1.27$$

Thus we see that the aggregate of $\hat{\beta}_1$ and $\hat{\beta}_2$ is not significantly different from a theorized value of -4.0.

8.2 Testing hypotheses concerning whole regression lines

In moving into considerations of multivariable equations we may require some mechanism for testing whether a whole line is significant in explaining a particular explanatory variable. To do this we begin by decomposing the variance in the explained variable Y_i in accordance with our familiar formula:

$$\begin{array}{ccccc} \text{total sum} & & \text{sum of squares} & & \text{sum of squares} \\ \text{of squares} & = & \text{explained by} & + & \text{unexplained by} \\ & & \text{regression} & & \text{regression} \\ \text{SST} & = & \text{SSR} & + & \text{SSE} \end{array}$$

An intuitive starting point in determining whether a whole regression line is significant in explaining the explained variable is to look at the ratio of the variation explained to that unexplained by the regression line. Thus our starting point is the quantity SSR/SSE. As the ratio of explained to unexplained variance increases, this quantity increases, pointing to the ability of our regression line. Should this quantity be low, then the amount of unexplained variance is large compared to the explained variance. This points to the inability of our regression line. Indeed, it can be shown that

$$\frac{\text{SSR}/(K-1)}{\text{SSE}/(n-K)} \overset{\mathrm{d}}{=} F_{K-1,\,n-K} \tag{8.3}$$

Verbally, this relationship advises that the ratio

$$\frac{\text{SSR}}{\text{SSE}} \cdot \frac{n-K}{K-1}$$

is distributed as an F distribution with $K-1, n-K$ degrees of freedom. In this relationship K equals the number of variables in the equation and n equals the sample size. The relationship also presents us with a tool for testing hypotheses concerning the whole line since both SSR and SSE are related to all the variables rather than individual ones. We can develop an analysis of variance table (table 8.1). Using our money-spent–percentage-vote example, we found that SST = 472.166 and SSE = 0.8513. Thus we find that SSR = 472.16 − 0.8513 = 471.31. Building an analysis of variance table we obtain table 8.2. In repairing to an F-table we see that this F value is extremely significant. Notice that in this bivariate case the F-value is the square of the t-value we obtained when we considered the individual regression-slope coefficient. The student should

Table 8.1

Source	degrees of freedom	SS	MS	F
Due to regression	$K-1$	SSR	$\dfrac{\text{SSR}}{K-1}$	$\dfrac{\text{SSR}/K-1}{\text{SSE}/n-K}$
Deviation from regression	$n-K$	SSE	$\dfrac{\text{SSE}}{n-K}$	
Total	$n-1$	SST	$\dfrac{\text{SST}}{n-1}$	

Table 8.2

Source	degrees of freedom	SS	MS	F
Due to regression	$2 - 1 = 1$	471.31	471.31	$\dfrac{471.31}{0.053} = 8930$
Deviation from regression	$18 - 2 = 16$	0.8513	0.053	
Total	17	472.16	27.877	

fully understand that this only happens when we deal with one explanatory variable.

We can rearrange the analysis of variance to form the following relationship using R^2 values only. We have shown earlier that $SSR/SST = R^2$ and can employ this to produce

$$F = \frac{R^2 \cdot (n - K)}{(1 - R^2)(K - 1)} \overset{d}{=} F_{K-1, \, n-K} \tag{8.4}$$

This will produce the same value of F as before.

Thus we see that the analysis of variance is really testing the goodness of fit of the line to the data. Whereas in the bivariate case the correlation coefficient tells us about the goodness of fit of the simple regression line, when we move into multivariate cases the analysis of variance table does a similar job but copes with all the explanatory variables simultaneously. The advantage of the analysis of variance is that it can accommodate both the bivariate and the multivariate cases. The analysis of variance tests the more extensive null-hypotheses,

$$H_0 \colon \hat{\beta}_1 = \hat{\beta}_2 = \hat{\beta}_3 = \cdots = \hat{\beta}_{K-1} = 0$$

while the t-test examines the more specific hypothesis,

$$H_0 \colon \hat{\beta}_k = 0 \quad \text{where } k = 1, 2, 3, \ldots, K - 1$$

It should be noted that if, in any multivariate regression, any one of the individual slope coefficients is significantly different from zero, then the F-statistic for the whole line will also be. However, if none of the individual slopes are significantly different from zero, there is still the chance that the whole line will explain a significant amount of variance of the explained variable. In this situation the effects of the individual explanatory variables are weak but they have a strong joint influence.

This is an indication of multicollinearity and provides us with a rule of thumb for determining whether multicollinearity exists.

When we obtain a situation where at the 5 percent level of significance the F-statistic for the line is significant while all the t-statistics for each individual regression slope are not significant, we declare that multicollinearity is too high and that we cannot disentangle the effects of the individual explanatory variables.

But this is a rule-of-thumb test for a whole data set. It may very well be the case that not all the explanatory variables are related. An important indicator in this situation is the correlation matrix. We can obtain the correlation of each explanatory variable with others. Inspection of this will highlight correlations close to an absolute value of unity.

The difficulty with this bivariate correlation is that it only points up collinearity, i.e., a dependence relationship between two variables. Multicollinearity, the condition when one explanatory variable is linearly related to some linear combination of the others, will not be detected. A rough and ready method here is to add each explanatory variable into the OLS regression in turn and to wait for the variance of the estimate to explode. This will give some idea of the culprit.

We can also produce the analysis of variance table in matrix form. From the previous chapter we saw that

$$\mathbf{Y'Y} - \frac{(\sum Y)^2}{n} = \boldsymbol{\beta'}\mathbf{X'Y} - \frac{(\sum Y)^2}{n} + \mathbf{e'e} \tag{8.5}$$

$$\text{(SST)} \qquad\qquad \text{(SSR)} \qquad \text{(SSE)}$$

Replacing the terms in the analysis of variance table by this matrix format we obtain table 8.3. Let us develop this for our contrived example. From

Table 8.3

Source	degrees of freedom	SS	MS	F
Due to regression	$K - 1$	$\boldsymbol{\hat{\beta}'}\mathbf{X'Y} - \dfrac{(\sum Y)^2}{n}$	$\dfrac{\left(\boldsymbol{\hat{\beta}'}\mathbf{X'Y} - \dfrac{(\sum Y)^2}{n}\right)}{K - 1}$	$\dfrac{\dfrac{\left(\boldsymbol{\hat{\beta}'}\mathbf{X'Y} - \dfrac{(\sum Y)^2}{n}\right)}{K - 1}}{\mathbf{e'e}/(n - K)}$
Deviation from regression	$n - K$	$\mathbf{e'e}$	$\mathbf{e'e}/(n - K)$	

the above table it is clear that we require the following matrix ingredients from the data matrices.

$$\hat{\beta}' = [4 \quad -1.45 \quad -3.10] \quad X' = \begin{bmatrix} 1 & 1 & 1 & 1 & 1 \\ 1 & -3 & 1 & 0 & 1 \\ -2 & 1 & -2 & 0 & 3 \end{bmatrix}$$

$$Y = \begin{bmatrix} 8 \\ 5 \\ 9 \\ 5 \\ -7 \end{bmatrix} \quad e = \begin{bmatrix} -0.75 \\ -0.25 \\ 0.25 \\ 1.00 \\ -0.25 \end{bmatrix}$$

From these we obtain

$$\hat{\beta}'X'Y = [4.0 \quad -1.45 \quad -3.10] \begin{bmatrix} 1 & 1 & 1 & 1 & 1 \\ 1 & -3 & 1 & 0 & 1 \\ -2 & 1 & -2 & 0 & 3 \end{bmatrix} \begin{bmatrix} 8 \\ 5 \\ 9 \\ 5 \\ -7 \end{bmatrix}$$

$$= 243.25$$

and

$$\frac{(\sum Y)^2}{n} = \frac{400}{5} = 80$$

thus

$$\hat{\beta}'X'Y - \frac{(\sum Y)^2}{n} = 163.25$$

$$e'e = [-0.75 \quad -0.25 \quad 0.25 \quad 1.0 \quad -0.25] \begin{bmatrix} -0.75 \\ -0.25 \\ 0.25 \\ 1.00 \\ -0.25 \end{bmatrix} = 1.75$$

The analysis of variance table becomes table 8.4. On consulting the F table we see that this value is highly significant, suggesting that the regression line as a whole significantly explains the explained variable.

Table 8.4

Source	degrees of freedom	SS	MS	F
Due to regression	$3 - 1$	163.25	$\dfrac{163.25}{2} = 81.625$	$\dfrac{81.625}{0.875} = 92.71$
Deviations from regression	$5 - 3$	1.75	$\dfrac{1.75}{2} = 0.875$	

Again this is little different from what was shown in Chapter 3, although the various ingredients for the calculations are obtained by way of the matrices generated from the multivariate situation. It has been included not only for the student to gain experience in handling the analysis of variance in matrix terms, but also because it allows some maneuvers most useful when comparing equations, as we shall see in Sections 8.2.2 and 8.2.3.

8.2.1 Presenting regression results

Most computer programs for regression produce the estimate for each regression slope, an estimate of the variance of the regression slope, an estimate of the intercept, and an F value for the whole regression line.

The regression slope and the variance of each regression slope allow tests concerning the influence of each of the explanatory variables on the explained variable. The same is also true of the intercept in the regression equation. The F value will determine whether the whole regression equation provides a significant explanation of the explained variable. All of this information is useful to the reader of research results. Some attempt should be made to present the information adequately. The following is suggested.

Suppose we have the regression model,

$$Y_i = \alpha + \beta_1 X_{i1} + \beta_2 X_{i2} + \beta_3 X_{i3} + u_i$$

After confrontation with the data, the following information is useful to the reader of the results.

$$Y = \underset{(S_0 = 1.38)}{3.2} + \underset{(S_1 = 1.14)}{1.15 X_1} + \underset{(S_2 = 1.87)}{2.5 X_2} + \underset{(S_3 = 2.07)}{7.2 X_3} \qquad \begin{aligned} R^2 &= 0.41 \\ n &= 31 \end{aligned}$$

where S_0, S_1, S_2, and S_3 are the standard error of the estimate. Let us try a whole testing procedure on this information. For each of the regression slopes and the intercept we shall test whether that particular estimate is significantly different from zero. We will take each individually. In each case we will adhere to a significance level of 0.01 one-tailed. The critical t-value for this significance level, taking into account the $n - K$ degrees of freedom (that is, $31 - 4 = 27$) is 2.473.

First test. In this test we shall be asking whether the intercept of the slope is different from zero.

$$H_0: \hat{\alpha} = 0 \quad H_A: \hat{\alpha} \neq 0$$

In developing the test statistic we divide the standard error of the estimate into the term $3.2 - 0$. This produces

$$t = \frac{3.2 - 0}{1.38} = 2.32$$

Since this value is less than the critical value, we accept the null-hypothesis and reject the alternative. We conclude that the intercept is not really significantly different from zero.

Second test. The hypothesis we shall be testing here is

$$H_0: \hat{\beta}_1 = 0 \quad H_A: \hat{\beta}_1 \neq 0$$

The test statistic is

$$t = \frac{1.15 - 0.0}{1.14} = 1.01$$

We again accept the null-hypothesis and reject the alternative.

Third test. This test concerns the estimate of the regression slope for X_2. The hypothesis here is

$$H_0: \hat{\beta}_2 = 0 \quad H_A: \hat{\beta}_2 \neq 0$$

and the test statistic for this is

$$t = \frac{2.5 - 0.0}{1.87} = 1.34$$

Again, we accept the null-hypothesis.

Fourth test. We can test the final regression slope in the equation. Our hypothesis is

$$H_0: \hat{\beta}_3 = 0 \quad H_A: \hat{\beta}_3 \neq 0$$

and the test statistic is

$$t = \frac{7.2 - 0.0}{2.07} = 3.47$$

Since this value is greater than our critical value for t, we reject the null-hypothesis and accept the alternative. We conclude that the variable X_3 has a significant influence on the explained variable Y.

Final test. So far we have tested whether each component of the right-hand side of the equation is different from zero. We found that only in the case of the variable X_3 was there any evidence that the explanatory variables accounted for a significant amount of the variance of the explained variable. It could well be the case that none of the explanatory variables individually accounts for much variance. Nevertheless, each explanatory variable might contribute an amount of explanation such that the total line does explain significantly. This is the purpose of the final test. It tells whether the line as a whole accounts for a significant proportion of the variance of the explained variable. Since we have already shown that X_3 alone does differ from zero, we can expect that the F statistic for the line will also prove significant. The hypothesis is

$$H_0: \hat{\beta}_1 = \hat{\beta}_2 = \hat{\beta}_3 = 0 \quad H_A: \hat{\beta}_1 = \hat{\beta}_2 = \hat{\beta}_3 \neq 0$$

and the test statistic is

$$F_{3, 27} = \frac{R^2}{1 - R^2} \cdot \frac{n - K}{K - 1} = \frac{0.41}{1 - 0.41} \cdot \frac{31 - 4}{4 - 1} = 6.25$$

The critical F value with these degrees of freedom is 4.60. Thus we see that the total regression line does explain a significant amount of variance in the explained variable.

8.2.2   Testing the explanatory power of added variables

Consider the situation where we have designed some regression equation but are unsure whether a particular variable or set of variables should be added. Thus we have the two competing regression equations:

$$Y_i = \beta_0 + \beta_1 X_{i1} + \beta_2 X_{i2} + \cdots + \beta_{R-1} X_{i, R-1}$$

and

$$Y_i = \beta_0 + \beta_1 X_{i1} + \beta_2 X_{i2} + \cdots + \beta_{R-1} X_{i, R-1}$$
$$+ \beta_R X_{iR} + \cdots + \beta_{Q-1} X_{i, Q-1}$$

In the first equation there are R variables, while in the second there are the same R variables *plus* some other variables, making a total of Q variables.

The hypothesis we test is

$$H_0: \hat{\beta}_R = \hat{\beta}_{R+1} = \hat{\beta}_{R+2} = \cdots = \hat{\beta}_{Q-1} = 0$$

To test this hypothesis we generate variance decomposition equations for both of the regression equations.

$$\text{SST} = \text{SSR}_R + \text{SSE}_R$$

$$\text{SST} = \text{SSR}_Q + \text{SSE}_Q$$

The first of these relationships refers to the first regression equation with R variables, and the second to the equation with Q variables. Notice that since we are dealing with the same explained variable, SST will be the same in both equations. If the added explanatory variables in the second equation do not provide any increased explanation, then SSR_R and SSR_Q will be the same. That is, the explained variance in the second equation will be the same as the explained variance in the first. This allows us to generate the following test statistic:

$$F = \frac{(\text{SSR}_Q - \text{SSR}_R)/\text{number of added variables}}{\text{SSE}_Q/(n - Q)}$$

$$\overset{\text{d}}{=} F \text{ number of added variables, } n - Q \tag{8.6}$$

The number of added variables is obviously equal to the number of variables in the second equation minus the number of variables in the first equation.

We have stated before that the addition of variables to any equation will not reduce the R^2 value. We see that

$$R_Q^2 \geqslant R_R^2$$

And since

$$R^2 = \frac{\text{SSR}}{\text{SST}}$$

we get

$$\frac{SSR_Q}{SST} \geqslant \frac{SSR_R}{SST}$$

Thus

$$SSR_Q \geqslant SSR_R$$

This informs us that we can never obtain a negative quantity in the test described above.

Let us return to our contrived example. We have the data:

Y_i	X_{i1}	X_{i2}
8	1	-2
5	-3	1
9	1	-2
5	0	0
-7	1	3

Suppose we have two competing theories. The first includes only X_1 whereas the second includes both X_1 and X_2. From the data above, the following estimating lines were obtained for the theories. With each of these estimating lines we also obtained R^2 values.

Theory 1 $\hat{Y}_i = 4.0 - 0.416X_{i1}$ $R^2 = 0.0127$,

$$n = 5, \quad K = 2$$

Theory 2 $\hat{Y}_i = 4.0 - 1.45X_{i1} - 3.10X_{i2}$ $R^2 = 0.989$,

$$n = 5, \quad K = 3$$

Returning to equation (8.6) and dividing top and bottom of the equation by the SST, we obtain the relationship

$$\frac{(R_Q{}^2 - R_R{}^2)/\text{number of added variables}}{(1 - R_Q{}^2)/(n - Q)} \overset{\mathrm{d}}{=} F \qquad \begin{matrix} \text{number of} \\ \text{added variables,} \\ n - Q \end{matrix} \quad (8.7)$$

In our example the subscript Q refers to the second theory and the subscript R to the first theory. Our test statistic becomes

$$\frac{(0.989 - 0.0172)/1}{(1 - 0.989)/(5 - 3)} = 178 \overset{\mathrm{d}}{=} F_{1,2}$$

This is obviously very significant, telling us that the addition of the second variable in theory 2 has a dramatic effect upon the amount of variance explained. Thus theory 2 is superior to theory 1 in terms of explaining the variance of the explained variable Y.

8.2.3   Testing the effects of added observations

There are certain research situations in which we are able to collect additional observations to those we may have already used to estimate regression parameters. In such fortunate situations we wish to test whether our new observations, let us say we have m of them, come from the same population as the previous n values. This may very well occur in time-series studies where one set of data was not available because the phenomenon had not yet occurred. At some point these data will become available.

Let us say that in the first part of this situation, where we have the original n observations, the estimating equation is

$$Y_i = \beta_0 + \beta_1 X_{i1} + \beta_2 X_{i2} + \cdots + \beta_{K-1} X_{i, K-1} \quad (i = 1, 2, 3, \ldots, n)$$

In the second part of the situation, where we have m observations, the estimating equation with these m observations alone is

$$Y_i = \gamma_0 + \gamma_1 X_{i1} + \gamma_2 X_{i2} + \cdots + \gamma_{K-1} X_{i, K-1}$$
$$(i = n + 1, n + 2, \ldots, n + m)$$

Further, the estimating equation using the whole data set of $m + n$ observations is

$$Y_i = \alpha_0 + \alpha_1 X_{i1} + \alpha_2 X_{i2} + \cdots + \alpha_{K-1} X_{i, K-1}$$
$$(i = 1, 2, 3, \ldots, m + n)$$

In this situation we are testing the following null-hypothesis:

$$H_0: \beta_0 = \gamma_0, \quad \beta_1 = \gamma_1, \quad \beta_2 = \gamma_2, \quad \ldots, \prime \beta_{K-1} = \gamma_{K-1}$$

and the test statistic is

$$F = \frac{(\mathrm{SSE}_C - \mathrm{SSE}_1 - \mathrm{SSE}_2)/K}{(\mathrm{SSE}_1 + \mathrm{SSE}_2)/(n + m - 2K)} \overset{d}{=} F_{K, n+m-2K} \tag{8.8}$$

In this relationship the subscript C refers to the combined data $n + m$, the subscript 1 refers to the original n observations, and the subscript 2 refers to the added m variables. The student will remember that

$$\text{SSE} = \sum (Y_i - \hat{Y}_i)^2 = \mathbf{e'e}$$

We can exploit the following example.

Initial data			Additional data		
Y_i	X_{i1}	X_{i2}	Y_i	X_{i1}	X_{i2}
8	1	-2	8	1	-2
5	-3	1	12	4	1
9	1	-2	10	2	-2
5	0	0	12	1	0
-7	1	3	4	0	3

In processing the original data we obtained the following estimating equation:

$$\hat{Y}_i = 4.00 - 1.45X_{i1} - 3.10X_{i2}$$

This produces the following residuals:

$$\mathbf{e}_1 = \begin{bmatrix} -0.75 \\ -0.25 \\ 0.25 \\ 1.00 \\ -0.25 \end{bmatrix}$$

In calculating the SSE_1, we get 1.75. Processing the additional data, we obtain the estimating equation

$$\hat{Y}_i = 7.20 + 1.0X_{i1} - 0.33X_{i2}$$

This produces the residual vector

$$\mathbf{e}_2 = \begin{bmatrix} 1.133 \\ -0.867 \\ -1.867 \\ 3.800 \\ -2.200 \end{bmatrix}$$

The SSE_2 for these residuals is 24.802. Adding both sets of data to form a combined set after regression produces

$$\hat{Y}_i = 6.005 + 0.595X_{i1} - 1.557X_{i2}$$

The following residuals are calculated from this regression:

$$\mathbf{e}_C = \begin{bmatrix} -1.714 \\ 2.338 \\ -0.714 \\ -1.005 \\ -8.929 \\ -1.714 \\ 5.171 \\ -1.500 \\ 5.400 \\ 2.666 \end{bmatrix}$$

Thus the SSE_C from this combined sample is 157.854. If we insert all of this in the test statistic (8.8) we obtain

$$\frac{(157.854 - 1.75 - 24.802)/3}{(1.75 + 24.802)/[5 + 5 - 2(3)]} = 6.53$$

With 3, 4 degrees of freedom this statistic is significant at the 0.05 level, indicating that the additional data set is significantly different from the original set.

There is one proviso in employing this technique. The number of additional observations must be at least equal to the number of variables in the estimating equation. Thus, formally,

$$m \geqslant K$$

8.3   Confidence intervals

In Chapter 3 we described various procedures for obtaining confidence intervals for the slope and intercept estimates and for $\hat{Y}_i$. We shall repeat such descriptions here using multivariate examples.

8.3.1   Confidence intervals for the regression slope and intercept

That the researcher is in a multivariate situation provides no difficulties in establishing confidence intervals for the regression estimates. As we observed in Chapter 3, a confidence interval is produced using the relationship:

$$P[\hat{\beta}_k - (t)\sqrt{\text{var}(\hat{\beta}_k)} < \beta_k < \hat{\beta}_k + (t)\sqrt{\text{var}(\hat{\beta}_k)}] = 1 - \lambda$$

In this equation k goes from 0 to $K - 1$. Thus to determine a confidence interval for any particular regression estimate, we need the estimate itself plus the variance of that estimate. The latter is obtained from the variance-covariance matrix for the regression. The student has already been instructed in getting these components. It then becomes a simple arithmetic process.

8.3.2   Confidence intervals for Y_i

Again referring to Chapter 3, we saw that the confidence interval for $\hat{Y}_i$ can be obtained from the relationship

$$P[\hat{Y}_i - (t)\sqrt{\text{var}(\hat{Y}_i)} < Y_i < \hat{Y}_i + (t)\sqrt{\text{var}(\hat{Y}_i)}] = 1 - \lambda$$

In this relationship the variance of $\hat{Y}_i$ will be different for each value of $\hat{Y}_i$. In obtaining our estimate of $\hat{Y}_i$, as we move towards the means of the distributions the variance of $\hat{Y}_i$ becomes smaller and the confidence interval obtained from it reduces. This is shown in diagram 3.7. In the multivariate situation a similar phenomenon occurs but in a many-dimensioned space.

For the relationship above we need two ingredients from our data; the estimate $\hat{Y}_i$ and the variance of this estimate. The former is obtained from the estimating equation as usual. The latter is obtained from the matrix relationship:

$$\text{var}(\hat{Y}_i) = \mathbf{X}_* (\text{var-cov matrix of regression estimates}) \mathbf{X}'_*$$
$$= \mathbf{X}_* [\sigma^2 (\mathbf{X}'\mathbf{X})^{-1}] \mathbf{X}'_* \tag{8.9}$$

In this equation $\mathbf{X}_*$ is a row vector of the values of each explanatory variable for the particular $\hat{Y}_i$ that we require. Thus we premultiply the variance-covariance matrix by this vector and then postmultiply all of this by the transpose of the vector. Notice that the variance-covariance matrix will be constant for a particular regression situation, but the $\mathbf{X}_*$ will change. Thus we see that the variance of $\hat{Y}_i$ will change according to the $\mathbf{X}_*$ vector, producing similar changes in the confidence interval for $\hat{Y}_i$.

Let me illustrate with an example. Using the data that follow,

Y_i	X_1	X_2
8	1	-2
5	-3	1
9	1	-2
5	0	0
-7	1	3

we have frequently obtained the estimating equation

$$\hat{Y}_i = 4.0 - 1.45X_{i1} - 3.10X_{i2}$$

I will develop two confidence intervals, one at the extreme of these distributions of the explanatory variables and one more towards the center of the distribution. In this way we shall see the change in the variance of Y_i and consequently in the confidence interval.

In our first example $\mathbf{X}_* = \begin{bmatrix} 1 & 1 & 3 \end{bmatrix}$. In the second example $\mathbf{X}_* = \begin{bmatrix} 1 & 0 & 0 \end{bmatrix}$. Although we only have two explanatory variables the value 1 is placed in the left-hand side of the row vector to accommodate the constant term. We have also seen that the variance-covariance matrix of the regression estimates in this equation is

$$\text{var-cov} \, (\hat{\beta}) = \begin{bmatrix} 0.175 & 0.0 & 0.0 \\ 0.0 & 0.078 & 0.0175 \\ 0.0 & 0.0175 & 0.0525 \end{bmatrix}$$

Thus we have all the ingredients we need for determining the confidence intervals, and we can take the first example.

Putting the $\mathbf{X}_*$ values into the estimating equation, we get

$$\hat{Y}_i = 4.0 - 1.45(1) - 3.10(3) = -6.75$$

To obtain the variance of Y_i we use

$$\text{var} \, (\hat{Y}_i) = \begin{bmatrix} 1 & 1 & 3 \end{bmatrix} \begin{bmatrix} 0.175 & 0 & 0 \\ 0 & 0.078 & 0.0175 \\ 0 & 0.0175 & 0.0525 \end{bmatrix} \begin{bmatrix} 1 \\ 1 \\ 3 \end{bmatrix} = 0.8305$$

Putting all of this into our confidence-interval equation, we get

$$P[-6.75 - 4.303\sqrt{0.8305} < Y_i < -6.75 + 4.303\sqrt{0.8305}] = 0.95$$

In this equation, 4.303 is the t value corresponding to the level of confidence; with $5 - 3$, i.e., $n - K$, degrees of freedom, the confidence interval becomes

$$P[-10.67 < Y_i < -2.83] = 0.95$$

Let us go on to the second example. We obtain $\hat{Y}_i$ thus:

$$\hat{Y}_i = 4.0 - 1.45(0) - 3.10(0) = 4.0$$

We obtain the variance of $\hat{Y}_i$ from

$$\text{var}(\hat{Y}_i) = \begin{bmatrix} 1 & 0 & 0 \end{bmatrix} \begin{bmatrix} 0.175 & 0 & 0 \\ 0 & 0.078 & 0.0175 \\ 0 & 0.0175 & 0.0525 \end{bmatrix} \begin{bmatrix} 1 \\ 0 \\ 0 \end{bmatrix} = 0.175$$

Notice the reduced variance at the center of the distribution. This will produce the following confidence interval for $\hat{Y}_i$:

$$P[4 - (4.303)\sqrt{0.175} < Y_i < 4 + (4.303)\sqrt{0.175}] = 0.95$$

where reduces on calculation to

$$P[2.19 < Y_i < 5.81] = 0.95$$

Thus we see that the intervals are smaller at the center of the explanatory variable distributions.

8.4   Prediction

As with the bivariate case, there may be research situations where we wish to predict some value of the explained variable, given a set of values for the explanatory variables. Such predictions may be inside or outside our data span. The former would require interpolation, the latter extrapolation. As we saw in Chapter 3, not only is it possible to do this, it is also possible to obtain a confidence interval of our prediction. As always we need various ingredients to put into the confidence-level relationship. In this equation the prediction value of explained variable is denoted by $\hat{Y}_0$.

$$P[\hat{Y}_0 - (t)\,\text{var}\,(\hat{Y}_0) < Y_0 < \hat{Y}_0 + (t)\,\text{var}\,(\hat{Y}_0)] = 1 - \lambda$$

Thus we require not only $\hat{Y}_0$ but also an estimate of its variance. $\hat{Y}_0$ is acquired by simply inserting the values of the explanatory variables. The variance is obtained by the following relationship:

$$\text{var}\,(\hat{Y}_0) = \text{var}\,(\hat{Y}_i) + \text{variance of the residuals} \qquad (8.10)$$

The reason for adding the variance of the residuals to the variance of $\hat{Y}_i$ is illustrated in diagram 3.9. The variance of $\hat{Y}_i$ is obtained as in Section 8.3.2. The variance of the residuals is obtained from the relationship

$$\text{var}\,(e_i) = \frac{e'e}{n - K}$$

Let us obtain a prediction for the $\mathbf{X}_* = [1\ 3\ 5]$. Notice that this data set is completely outside our observations.

Y_i	X_1	X_2
8	1	-2
5	-3	1
9	1	-2
5	0	0
-7	1	3

Our $\hat{Y}_0$ is given by

$$\hat{Y}_0 = 4.0 - 1.45(3) - 3.10(5) = -15.85$$

and variance of residuals is

$$\frac{1.75}{5 - 3}$$

The variance of $\hat{Y}_0$ is given by

$$\text{var}\,(\hat{Y}_0) = [1\ \ 3\ \ 5] \begin{bmatrix} 0.175 & 0 & 0 \\ 0 & 0.078 & 0.0175 \\ 0 & 0.0175 & 0.0525 \end{bmatrix} \begin{bmatrix} 1 \\ 3 \\ 5 \end{bmatrix} + \frac{1.75}{5 - 3}$$

$$= 3.5895$$

Thus the confidence intervals are obtained by the relationship

$$P[-15.85 - 4.303\sqrt{3.5895} < Y_0 < -15.85 + 4.303\sqrt{3.5895}] = 0.95$$

This reduces on calculation to

$$P[-23.98 < Y_0 < -7.72] = 0.95$$

This is obviously a wide interval because we are predicting well outside our observation set. As we move to predictions within the data set, the interval will get smaller and we can be more confident of our prediction. This accords with common sense.

We have now completed our testing in the multivariate situation. We have used the same small examples all the way through in an attempt to encourage the student to try the matrix calculations.

8.5   Misspecification of regression models

There is a plethora of ways in which the researcher can misspecify the regression model. These range from including irrelevant variables and excluding relevant variables to errors in assuming a linear relationship between explanatory variables. In this section we shall deal with the former exclusively, that is, situations where we have included irrelevant variables or excluded relevant variables.

Let us take the more dangerous situation first: the omission of a relevant variable. Suppose we suggest the following model in some research situation:

$$\mathbf{Y} = \mathbf{X}_1\mathbf{B}_1$$

In this model the $\mathbf{X}_1$ matrix will refer to a particular set of explanatory variables and the $\mathbf{B}_1$ matrix to the regression coefficients for these explanatory variables.

We have seen that

$$\hat{\mathbf{B}}_1 = (\mathbf{X}_1'\mathbf{X}_1)^{-1}\mathbf{X}_1'\mathbf{Y}$$

Replacing $\mathbf{Y}$ by the $\mathbf{X}_1\mathbf{B}_1$ we obtain

$$\hat{\mathbf{B}}_1 = (\mathbf{X}_1'\mathbf{X}_1)^{-1}\mathbf{X}_1'\mathbf{X}_1\mathbf{B}_1$$

Suppose, however, that in generating this model we have omitted some relevant variables and that the true model is

$$Y = X_1 B_1 + X_2 B_2 \tag{8.11}$$

In this model the explanatory variables X_2 and the regression coefficients B_2 are those omitted from the model. So our calculation of B_1 would follow the steps

$$\hat{B}_1 = (X_1' X_1)^{-1} X_1' Y$$

And substituting Y we get

$$\hat{B}_1 = (X_1' X_1)^{-1} X_1' (X_1 B_1 + X_2 B_2)$$
$$= B_1 + (X_1' X_1)^{-1} X_1' X_2 B_2$$

Our estimate of $\hat{B}_1$ will be biased by the quantity $(X_1' X_1)^{-1} X_1' X_2 B_2$ if we omit the relevant variables X_2. Examining this bias we see that if B_2 is zero then the bias would disappear. Looking at the quantity $X_1' X_2$ we see that if this equals zero then the bias will also be zero. However, the quantity $X_1' X_2$ represents the covariance of the variables in the matrices. Thus we see that if the explanatory variables included in the regression do not covary with those left out of the model, we do not bias the estimates obtained in the incorrect model.

We can make the following assertions concerning the effects of omitting relevant variables from our regression:

(1) If variables in the model covary with those left out of the model, then all the regression estimates will be biased. Similarly, the variance-covariance matrix for the regression estimate will be biased upwards, making it increasingly difficult to reject null-hypotheses.

(2) If the variables in the model do not covary with those left out of the model, then the estimates of the regression slopes are unbiased, although the variance-covariance matrix of the regression slope estimates will still be biased upwards, providing less chance of rejecting the null-hypothesis. The intercept will be biased unless the mean of the explanatory variables left out of the model is zero. This can be shown by pointing out

that the variables left out of the model will find their way into the error term. The error term will act normally because it is uncorrelated with the explanatory variable. Nevertheless, the expected value of the error term will not be zero unless all the distributions that make it up have a mean of zero.

We can now consider the difficulties generated by including irrelevant variables in our model. If we take equation (8.11) we can consider the matrix X_2 as those variables we have included in the model but which are irrelevant. Given that they are irrelevant, we know that B_2 will be zero. But more importantly, we want to gauge the effect of inserting these irrelevant variables on the estimates of B_1, the regression coefficients for the explanatory variables that should be in the regression model. The estimate for B_1 will be given by the equation

$$\hat{B}_1 = B_1 + (X_1'X_1)^{-1}X_1'X_2B_2$$

But we know that $B_2 = 0$ and thus the whole bias reduces to zero and our equation becomes

$$\hat{B}_1 = B_1$$

We can conclude that the inclusion of irrelevant variables in our regression equations does not bias the estimates of the relevant regression coefficients. In essence, all that will happen is that the regression estimates for the irrelevant variables will be zero or close to it. Unfortunately, the addition of irrelevant variables affects our degrees of freedom. So we do pay some price.

Summarizing our findings from this section, we see that the exclusion of relevant variables from the regression model can have serious effects upon the accuracy of the estimates included in the model and will certainly render inaccurate our inference-testing procedures. The difficulty is that it is very hard to determine when one has omitted a relevant variable. In contrast, the inclusion of irrelevant variables has no detrimental effect on the process. If in doubt whether to include any variable in a regression analysis, we should go ahead and do so. Although this may seem atheoretical, it is far better than obtaining inaccurate inferences, keeping in mind at all times the unfortunate effect upon our degrees of freedom.

Further readings

Substantive
Gray, V. 1973. "Innovation in the States: A Diffusion Study." *Am. Pol. Sci. Review* 67: 1174.
Kort, F. 1968. "A Nonlinear Model for the Analysis of Judicial Decisions." *Am. Pol. Sci. Review* 62: 546.
Kramer, G. H. 1971. "Short-Term Fluctuations in U.S. Voting Behavior, 1896–1964." *Am. Pol. Sci. Review* 65: 131.

Statistical
Goldberger: pp. 163–68.
Johnston: pp. 135–55.
Kmenta: pp. 216–46.

9

other regression models

In this final chapter of Part Two we should discuss variants of the classical regression procedure. There are many research situations where the classical linear model is not really applicable. Regression extensions have been developed to be useful in these situations. In this chapter we will discuss

(1) binary or dummy variables,
(2) nonlinear models,
(3) restricted coefficients.

In addition there are other situations where the classical regression technique is applicable but needs careful operation with full knowledge of difficulties. Such situations are also included in this catch-all chapter. Under this we will discuss

(4) lagged variables,
(5) stepwise regression,
(6) missing data.

Because this is a catch-all chapter the student should not necessarily expect continuity from one section to another. Each section can be read or used independently.

9.1 Binary or dummy variables

There are basically two ways that we conclude we have some variable with discrete categories. The first is by looking at a scatterplot of data; the second is by determining the substantive nature of variables. In the first situation it becomes obvious from the scatterplot that some hitherto unknown variable is producing patterning in our data.

Take the plot of defense expenditure over the years 1935 to 1950 (diagram 9.1). It can readily be seen that, during the period 1942–1945, there is some step increase in the level of defense expenditure. One may conclude that the war would account for this. Thus we can insert into our research situation a binary variable, i.e., a two-state variable—one state being peace, the other war.

Diagram 9.1

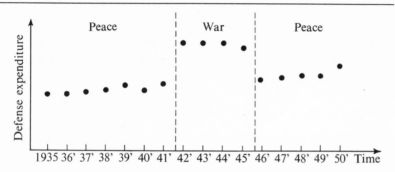

But some variables are binary by nature. Someone is either black or nonblack. There are communist and noncommunist countries. You are either male or female. In essence, we can distinguish three kinds of such binary variables. These are:

(1) time effects,
(2) spatial effects,
(3) qualitative variables.

The example of the defense expenditures would be a binary effect under the time heading. Under the spatial heading one could consider shifts in mean levels of certain political variables as one moves from one part of the international system to another. For example one would expect economic development to be substantially different in Third World countries as opposed to non–Third World countries. Qualitative variables such as black-white, communist-noncommunist are also found in political research situations.

Finally, it is also possible to generate binary variables out of variables that are continuous. For example, we can divide people up into young and old, although, of course, we could easily process their exact age. Similarly, we can divide nations up into rich and poor, although we may have quite accurate data on their actual GNP.

Since such discrete-state variables exist in political science, it becomes necessary to develop a manner of incorporating them in our regression models. In doing this we will also reduce some of the inaccuracy in our regression estimating. Consider the scattergram in diagram 9.2. The regression line A, which would be the usual least-squares regression line, goes through none of the actual data points. It is inadequate in explaining either the data points produced in peace or those produced in war. The estimated regression line is biased, and the variance of the regression estimates will also be inaccurate. But the lines B and C individually explain the war years and the peace years respectively. There are two regression lines. The aim of using dummy variables is to produce a regression relationship in which one equation accounts for both the peace and war conditions.

Diagram 9.2

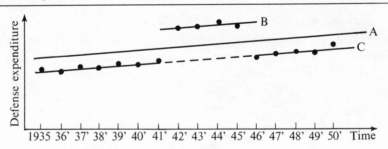

9.1.1 Quantitative explained variables—qualitative explanatory variables

Let us first consider situations where there is only one explanatory variable and it is binary. In the situation shown below we are interested in predicting people's salaries and whether skin color has any effect upon this.

As diagram 9.3 shows, the mean level of salary for whites is μ_W, while for nonwhites the mean salary is μ_{NW}. Thus we need two equations to describe the data on this scattergram:

$$Y_i = \mu_W + e_i \quad \text{for white salaries}$$
$$Y_i = \mu_{NW} + e_i \quad \text{for nonwhite salaries}$$

But if we produce a dummy variable X_i such that

$$X_i = 0 \quad \text{when the salary receiver is nonwhite}$$
$$X_i = 1 \quad \text{when the salary receiver is white}$$

then these two equations can be covered by the one regression equation,

$$Y_i = \beta X_i + \alpha + e_i$$

In this equation, α, the regression intercept, is equal to the mean salary of the nonwhites. The regression slope β is some measure of the effects on salary of being white. If this slope is zero, we can say that color has no effect on salary level, and the level of salary reduces to α which is the level of nonwhites. Thus β is the difference between the mean levels of the two groups and is equal to $\mu_W - \mu_{NW}$. Testing whether β is different from zero is determining if there is evidence that the color of a person's skin affects level of salary.

Diagram 9.3

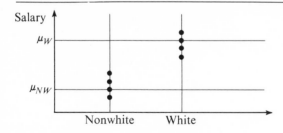

The researcher carries out simple regression on the data matrix example shown in table 9.1. Notice that we did not construct two variables to account for the categories of white and nonwhite. That is, we did not let

$$X_{i1} = 1 \text{ for a white}$$
$$X_{i2} = 1 \text{ for a nonwhite}$$

Table 9.1

	Y_i	X_i	Remarks
Case 1	$12,000	1	white person
Case 2	$12,500	1	white person
Case 3	$ 9,000	0	nonwhite person
Case 4	$11,200	1	white person
.	.	.	.
Case n	$ 7,000	0	nonwhite person

Had we done this, we would have produced the data matrix in table 9.2. In this matrix the two right-hand columns are correlated—if you are not nonwhite then you are white. One column is a perfect predictor of the other. And as we saw in Chapters 6 and 7, such situations render the $X'X$ matrix as singular and incapable of being inverted. Thus we have to stipulate the following condition for this procedure. If there are G states of any variable, we can produce $G - 1$ dummy variables in the regression equation. Since we have two states to our variable, white and nonwhite, we are prohibited from having more than $2 - 1 = 1$ dummy variables.

We can process discrete variables with more than two categories. For instance, if we increase our categories to distinguish between blacks and nonwhites, we would have produced a three-state variable; white, black and others. We can represent these three states by two $(3 - 1)$ dummy variables. Let these be

X_{i1} for a white person

X_{i2} for a black person

By doing this we reduce the following three equations:

$Y_i = \mu_W + e_i$ for whites, μ_W is white mean salary

$Y_i = \mu_B + e_i$ for blacks, μ_B is black mean salary

Table 9.2

	Y_i	X_{i1}	X_{i2}
Case 1	$12,000	1	0
Case 2	$12,500	1	0
Case 3	$ 9,000	0	1
Case 4	$11,200	1	0
.	.	.	.
Case n	$ 7,000	0	1

$$Y_i = \mu_O + e_i \quad \text{for other, } \mu_O \text{ is others' mean salary}$$

to the single equation

$$Y_i = \alpha + \beta_1 X_{i1} + \beta_2 X_{i2} + e_i$$

In this equation, if $\beta_1 = 0$, then being white has no effect on salary level; and if $\beta_2 = 0$, then being black has no effect on salary level. When both are zero, all salaries are the same; that is, they are constant at α.

We see therefore that a number of states of the same variable can be accommodated, but in addition, we can cope with more than one variable. Let us suppose that we are interested in education as well as skin color as a predictor of salary level. Two possible states of education might be Ph.D. and non-Ph.D. To sketch this we have to resort to a three-dimensional scattergram (diagram 9.4). To represent this scattergram mathematically we would require six equations.

$$Y_i = \mu_{W, P} + e_i \quad \text{where } \mu_{W, P} = \text{mean salary for a white Ph.D.}$$
$$Y_i = \mu_{W, N} + e_i \quad \text{where } \mu_{W, N} = \text{mean salary for a white non-Ph.D.}$$
$$Y_i = \mu_{B, P} + e_i \quad \text{where } \mu_{B, P} = \text{mean salary for a black Ph.D.}$$
$$Y_i = \mu_{B, N} + e_i \quad \text{where } \mu_{B, N} = \text{mean salary for a black non-Ph.D.}$$
$$Y_i = \mu_{O, P} + e_i \quad \text{where } \mu_{O, P} = \text{mean salary for others with a Ph.D.}$$
$$Y_i = \mu_{O, N} + e_i \quad \text{where } \mu_{O, N} = \text{mean salary for others without a Ph.D.}$$

By using the new dummy variable Z so that

$$Z_i = 0 \quad \text{for non-Ph.D.'s}$$
$$Z_i = 1 \quad \text{for Ph.D.'s}$$

Diagram 9.4

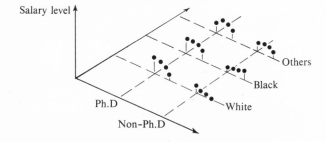

we can reduce these six equations to one regression equation,

$$Y_i = \alpha + \beta_1 X_{i1} + \beta_2 X_{i2} + \gamma Z_i + e_i$$

In this equation, if none of the regression slopes are different from zero then neither color nor education has any effect upon salary. If γ is zero, then having a Ph.D. does not have an effect. We saw in the previous situation the meaning of either β_1 or β_2 being equal to zero.

At this point it might have occurred to the student that this technique produces the same results as would one-way analysis of variance in the case of both skin color and education. Thus the use of dummy variables equates with classical analysis of variance.

We can go even further with this procedure and consider the interaction between variables. For instance, although the idea is far-fetched, suppose that neither having a Ph.D. nor skin color alone has an effect on salary, but that a combination of the two does. Skin color and education interact. These are additional variables and are produced easily, as table 9.3 shows.

Table 9.3

	Y_i	X_{i1}	X_{i2}	Z_i	$X_{i1}Z_i$	$X_{i2}Z_i$
Case 1	$12,000	1	0	1	1	0
Case 2	$12,500	1	0	0	0	0
Case 3	$ 9,000	0	0	1	0	0
Case 4	$11,200	1	0	0	0	0
.	.	.	.	.	.	.
Case n	$ 7,000	0	1	1	0	1

The regression equation for this would be

$$Y_i = \alpha + \beta_1 X_{i1} + \beta_2 X_{i2} + \gamma Z_i + \delta_1 X_{i1}Z_i + \delta_2 X_{i2}Z_i + e_i$$

where the estimates δ_1 and δ_2 measure the interaction effects of education and skin color. If δ_1 or δ_2 is zero, the researcher infers that there is no interaction effect between the two variables. Notice in the data matrix how the column $X_{i1}Z_i$ is obtained by multiplying column X_{i1} by column Z_i. Column $X_{i2}Z_i$ is produced by multiplying column X_{i2} by column Z_i.

9.1.2　Quantitative explained variables—qualitative and quantitative explanatory variables

In reality, of course, the explanatory variables are never just quantitative but a mixture of qualitative and quantitative. For instance, consider relating defense expenditure to a nation's total tax income over the period 1935–1950, as in diagram 9.5. As the diagram shows, the total tax input to the system grew steadily, but the defense expenditure exhibited a step increase and decrease. Further examination shows that these changes were in the war years. The regression line going through the data points for the war years is different from that going through the data points in the nonwar years. Obviously, if we do not include some binary variable to take care of the step in the data points, our regression estimates will be neither use nor ornament.

Diagram 9.5

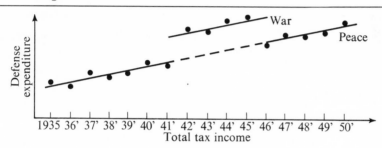

Instead of processing the regression model,

$$D_t = \alpha + \beta T_t + e_t$$

which would try to locate the line in between the data points, we process the regression model,

$$D_t = \alpha + \beta T_t + \gamma Z_t + e_t$$

where Z_t is a dummy variable such that when

$Z_t = 0$ it is peacetime

$Z_t = 1$ it is wartime

In essence what we are saying is that in peacetime the intercept is α, but in wartime the intercept is $\alpha + \gamma$. Thus we can infer that wartime changes the intercept value of the slope only. Diagram 9.6 shows this. The regression lines are parallel, indicating that the slopes are equal. The addition of the dummy variable has helped us cope with a situation where the intercept of the line changes. We can also exploit a dummy variable to cope with the data situation in diagram 9.7. With this data pattern it is

Diagram 9.6

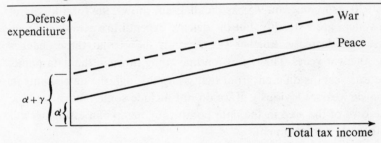

the slope of the regression line that appears to have been changed by the war period. We see that regression line A has a slope different from that of the regression line for the period of peace, B. For line B we would require the regression line

$$D_t = \alpha + \beta_1 T_t + e_t$$

The regression line for line A would be

$$D_t = \alpha + \beta_2 T_t + e_t$$

But using the dummy variable Z_i where Z_i equals zero in peacetime and unity in wartime, we can incorporate both lines in the same regression model.

$$D_t = \alpha + \beta_1 T_t + \gamma_1 Z_t T_t + e_i$$

In this equation, when Z_i is zero in peacetime the equation reduces to that for line B in diagram 9.7. In wartime the equation becomes

$$D_t = \alpha + \beta_1 T_t + \gamma_1 T_t + e_t$$

Diagram 9.7

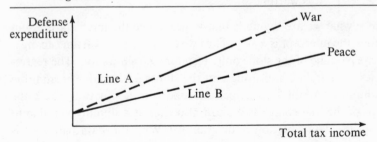

Diagram 9.8

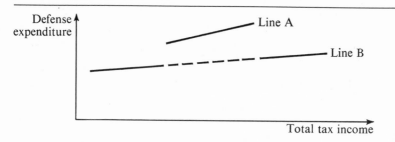

which reduces to

$$D_t = \alpha + (\beta_1 + \gamma_1)T_t + e_t$$

Thus $(\beta_1 + \gamma_1)$ equals β_2 in the regression equation for line A. In this case the intercept value does not change.

Obviously we can combine both of the situations above to produce a regression equation to manage changes in both the slope and intercept shown in diagram 9.8. The regression equation for line A would be

$$D_t = \alpha_1 + \beta_1 T_t + e_t$$

and for line B,

$$D_t = \alpha_2 + \beta_2 T_t + e_t$$

Using the same dummy variable as in the other cases, we can prepare one equation to take care of the whole data set.

$$D_t = \alpha_2 + \gamma_1 Z_t + \beta_2 T_t + \gamma_2 Z_t T_t + e_t$$

When Z_t equals 0, the equation reduces to that for line B. When Z_t is 1 the equation becomes

$$D_t = (\alpha_1 + \gamma_1) + (\beta_2 + \gamma_2)T_t + e_t$$

where $(\alpha_2 + \gamma_1)$ and $(\beta_2 + \gamma_2)$ are equal to α_1 and β_1 respectively from the equation for line A.

So we see that dummy variables can be exploited to cope with many difficult research situations. It can also be shown that the quasi-experimental ideas expressed by Campbell[1] are in essence a variation of the dummy variable technique.

1. D. Campbell and J. Stanley, *Experimental and Quasi-Experimental Designs for Research* (Chicago, Ill.: Rand-McNally, 1966).

9.1.3 Qualitative explained variable—quantitative explanatory variables

There will be research situations where the explained variable is polychotomous while the explanatory variables are quantitative. For example, we may have a situation in which we want to model the effect of some voter's salary on that person's voting behavior. Since voting behavior is manifested in a vote for either the Democrats or the Republicans, we see that the explained variable is dichotomous. Obviously the explanatory variable is quantitative. Our regression equation will be

$$Y_i = \beta_0 + \beta_1 X_{i1} + u_i$$

In this equation Y_i represents the way any individual votes and will be, say, 0 for a Democratic vote and 1 for a Republican vote. Variable X_1 is the individual's salary.

The estimating equation for this model is

$$\hat{Y}_i = \hat{\beta}_0 + \hat{\beta}_1 X_{i1}$$

It is important that the student understands the nature of $\hat{Y}_i$ in this estimating equation. Obviously, since any regression line is continuous, the value of $\hat{Y}_i$ will not always fall exactly on 0 or 1. Thus we have to interpret $\hat{Y}_i$ as a probability. Given any salary level, the estimated value $\hat{Y}_i$ this produces in the estimating equation is taken as the probability that the individual will vote Republican. Alternatively, $1 - \hat{Y}_i$ is the probability that the individual will vote Democratic. We assume that all people will vote.

However, there are three major problems with this regression situation. First, since $\hat{Y}_i$ can only achieve two values, 0 and 1, u_i also can only achieve two values, $-\beta_0 - \beta_1 X_{i1}$ and $1 - \beta_0 - \beta_1 X_{i1}$. This means that the error is dependent upon the explained variable, giving us a heteroscedastic difficulty. As we have seen before, the resulting estimates will be unbiased but inefficient. A second problem also derives from the distribution of the error terms. Since the distribution is demonstrably not normal, the various hypothesis-testing procedures are not really applicable, rendering it unwise to make any other but cautionary noises about one's findings.

The final problem concerns values of Y_i which fall below 0 and above 1. Since all probability statements fall within the interval 0 to 1, values of Y_i outside this interval are nonsensical. But such values may very well occur since our regression line is of infinite length. To overcome this difficulty we observe the following decision rule. Letting $\hat{\hat{Y}}_i$ be an ascribed value, we determine according to the following rules:

$$\hat{\hat{Y}}_i = \hat{Y}_i \quad \text{if} \quad 0 \leqslant \hat{Y}_i \leqslant 1$$
$$\hat{\hat{Y}}_i = 0 \quad \text{if} \quad \hat{Y}_i \leqslant 0$$
$$\hat{\hat{Y}}_i = 1 \quad \text{if} \quad \hat{Y}_i \geqslant 1$$

Diagram 9.9 shows the effect of this decision rule.

Diagram 9.9

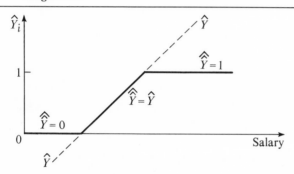

9.1.4 Qualitative explained variables—qualitative explanatory variables

A final research situation using qualitative data is when both the explained and explanatory variables are qualitative. For example, we might again want to model an individual's voting preference, given that we know the individual's class background and education. In the example we shall be employing later in this section, we allow the voting preference to be dichotomous, i.e., Democratic or Republican, the class variable to be tri-chotomous, i.e., lower, middle, and upper class, and the education variable to be dichotomous, i.e., Ph.D. and non-Ph.D.

Kort[2] has described with unusual clarity a technique for developing what he calls discriminant functions using ordinary multiple regression. Let us proceed with our example by way of an explanation of the process. In this example we allow Y_i to have only two values; 0 when the individual votes Democratic and 1 when the individual votes Republican. One explanatory variable is X_1, which represents the class variable and achieves

2. F. Kort, "Regression Analysis and Discriminant Analysis," *American Political Science Review* 67 (1973): 555–59.

−1 when the individual is from the lower class, 0 when the individual is from the middle class, and 1 when the individual is from the upper class. The other explanatory variable, education, achieves two values; 0 for the individual with a Ph.D. and 1 for the individual without a Ph.D.

The regression model for this would be

$$Y_i = \beta_0 + \beta_1 X_{i1} + \beta_2 X_{i2} + u_i$$

The discriminant function for the model would be

$$Z_i = c_1 X_{i1} + c_2 X_{i2}$$

The relationship between, say, β_1 and c_1 is given by

$$c_1 = \beta_1 \frac{n}{n_1 n_2}$$

where n = total sample size
n_1 = number of sample with a 0 explained variable
n_2 = number of sample with a 1 explained variable

Each conversion of β to c is done with multiplication by the ratio $n/n_1 n_2$. Thus we can use the easily obtained multiple regression estimates to obtain the discriminant function Z_i. This discriminant function should not be considered as an explained variable but an index. Kort also argues that the F value for the regression line can indicate whether a particular model is statistically significant or not and that such statistical significance passes over into our discriminant function. Thus if our regression line proves to be significant, we can assume that the discriminant function is also significant.

Let us go through the whole process, using data. The data are in table 9.4. On performing ordinary multiple regression on these data we produce the estimating equation

$$\hat{Y}_i = 0.29825 + 0.49123 X_{i1} + 0.15789 X_{i2}$$

The F value for this line is 14.79, which is significant far beyond the 0.01 level. Conversion to the discriminant function produces

$$\hat{Z}_i = 0.49123 \times \frac{20}{10 \times 10} X_{i1} + 0.15789 \times \frac{20}{10 \times 10} X_{i2}$$

$$\hat{Z}_i = 0.098 X_{i1} + 0.0316 X_{i2}$$

Table 9.4

Y	X_1	X_2	Y	X_1	X_2
1	1	1	0	−1	0
1	0	0	0	−1	0
1	1	1	0	0	0
1	1	0	0	0	0
1	1	1	0	−1	0
1	1	1	0	0	1
1	1	1	0	0	0
1	1	1	0	0	0
1	1	1	0	0	0
1	0	1	0	0	0

We are now in a position to produce a table of estimated values (table 9.5). If we want to make a prediction about the way an individual will vote, given that we know which class the individual comes from and whether the individual has a Ph.D., we need some decision rule regarding the value of the discriminant function. Kort suggests that if we take the mean of the Z_i values for the cases voting Republican, i.e., 0.0987, and the mean of the Z_i values for those voting Democratic, i.e., −0.0155, and then average these two, we end up with a cut-off value, i.e., 0.0416. If the estimated value for a particular individual is below this value, we predict that the individual will vote Democratic. Should the estimated value be above this 0.0416, we predict the individual will vote Republican. And indeed, looking at the data, our predictions would have been wrong only twice; case 2 and case 10.

Table 9.5

i	Y_i	$\hat{Y}_i$	$\hat{Z}_i$	i	Y_i	$\hat{Y}_i$	$\hat{Z}_i$
1	1	0.947	0.1234	11	0	−0.193	−0.0602
2	1	0.298	0.0000	12	0	−0.193	−0.0602
3	1	0.947	0.1234	13	0	0.298	0.0000
4	1	0.789	0.0918	14	0	0.298	0.0000
5	1	0.947	0.1234	15	0	−0.193	−0.0980
6	1	0.947	0.1234	16	0	0.456	0.0316
7	1	0.947	0.1234	17	0	0.298	0.0000
8	1	0.947	0.1234	18	0	0.298	0.0000
9	1	0.947	0.1234	19	0	0.456	0.0316
10	1	0.456	0.0316	20	0	0.298	0.0000

One sees therefore that the process is relatively simple, exploiting our already familiar regression programs. Again the student should be fore-warned that we have heteroscedasticity and nonnormal error distribution. It is because of this that I have some reservations about the statistical significance assertion that Kort makes. Nevertheless the technique is useful and as such is essential in the quiver of the politometrician. Added to this, there are the slightly more sophisticated *logit* and *probit* analyses, descriptions of which can be found elsewhere.[3]

9.2   Models that are nonlinear

We have confined all of our arguments so far to linear models. Un-fortunately, many of the relationships between variables are nonlinear. Unless we can in some way rearrange or transform these to linear models, there is no way that we can estimate their parameters using techniques so far described.

One can distinguish between *intrinsically linear* and *intrinsically non-linear* models. Intrinsically linear models can be transformed into linear models although they may be nonlinear. In many cases intrinsically nonlinear models cannot be so transformed because of difficulties with the error term.

Basically there are two indicators that a nonlinear model is appropriate. First, we may have theoretical indications of nonlinearity. For instance, an example we shall discuss more fully later concerns the growth of police expenditure in Chicago. Traditionally, growth models are of the form of compound-interest models or steady-growth models. Thus the researcher would go to one of these first. The second indicator of nonlinearity in models can come from a scattergram plot of the data. Any plot which seems to produce a form other than linear might require a nonlinear model. Some possible data plots are shown in diagram 9.10.

Intrinsically linear models can be broken down further into those models which are nonlinear in the variables, and those which are non-linear in the parameters. An example of the former would be

$$Y_1 = \alpha + \beta_1 X_i + \beta_2 X_i^2 + u_i$$

3. Goldberger, *Econometric Theory*, pp. 250 ff.

Diagram 9.10

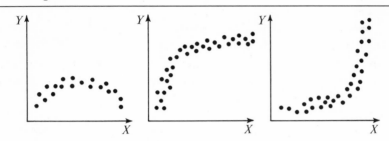

An example of the latter would be

$$Y_i = P_0(1 + i)^n$$

Let us first discuss the models that are nonlinear in variables.

9.2.1   Models linear in parameters, nonlinear in variables

There are two basic kinds of such models: polynomial models and reciprocal models. Perhaps the best first step in applying any model to data is
to explore polynomial models. In such models we add increasing powers
of the same variable. For example, a polynomial in X_i might be

$$Y_i = \alpha + \beta_1 X_{i1} + \beta_2 X_{i1}^2 + \beta_3 X_{i1}^3 + u_i$$

This type of model is the simplest to transform into a linear model capable
of regression by the techniques already discussed. All that is required is
to replace X_{i1}^2 and X_{i1}^3 by surrogates such as W and Z, where $W_{i2} = X_{i1}^2$
and $Z_{i3} = X_{i1}^3$. The regression equation becomes:

$$Y_i = \alpha + \beta_1 X_{i1} + \beta_2 W_{i2} + \beta_3 Z_{i3} + u_i$$

which is linear in X, W, and Z.

The use of the polynomial prior to any other model can be argued
because a polynomial can be constructed to fit any set of points. Any two
points can be described by a first-order polynomial, that is, a line; any
three points can be described by a second-order polynomial, and so on.
Thus any n points can be described by an $n - 1$ order polynomial. But
it should be remembered that constructing polynomials can be theoretically inefficient. A modest data set of sample size 25 could presumably be
best estimated by a 24th-order polynomial. Hardly parsimonious! In
addition, since the terms of the polynomial are some variable to increasing powers, all of the explanatory variables will be correlated, leading

to the most distressing multicollinearity problems and the consequent explosion of the variance-covariance matrix.

One can, of course, add higher powers of the polynomial and determine whether the added term increases significantly the amount of variance explained. In this way some semblance of parsimony might be retained. But all in all, the dangers in this technique of acting atheoretically must be balanced against the joy of just fitting some regression shape to the data.

The student should notice that in this transformation the error term was not changed in any way. Thus, assumptions made concerning the error terms also are not tampered with.

A second common nonlinear model is the reciprocal model. This is normally of the form

$$Y_i = \alpha + \beta \frac{1}{X_i} + u_i$$

This model is particularly useful when the data points appear to indicate that the explained variable approaches some limiting value, as in diagram 9.11. Again, the model is rendered processable by the simple provision of a surrogate. Thus letting $W_i = 1/X_i$ provides the model

$$Y_i = \alpha + \beta W_i + u_i$$

This is instantly recognizable as a linear regression formulation. The student should again notice that the error term is unaffected by the transformation.

There are many combinations and permutations of these two types and models, many of which can be extremely useful. And of course there are

Diagram 9.11

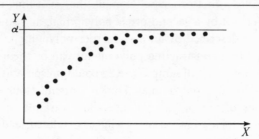

other functions which are applicable, so many that it would not be sensible to try to incorporate them here. There are special sources to which the student can go for further information.

9.2.2 Models nonlinear in parameters

The second kind of intrinsically linear model we should discuss is that which is nonlinear in the parameters.

For example, a colleague of mine has been trying to establish the growth rate of police cost in Chicago from the nineteenth century to the present. His data are police expenditures per year. Let us call the amount of money expended in the first year P_0. Each year, owing to inflation and other factors, there is more or less a steady growth increase which we can call i. After the first year, police expenditure would be

$$P_1 = P_0(1 + i)$$

And in the second year the expenditure would be given by

$$P_2 = P_1(1 + i) = P_0(1 + i)^2$$

Thus we see that a general model for the growth of expenditure will be

$$P_t = P_0(1 + i)^t \quad \text{where } t = \text{year number}$$

This model is hardly linear but perhaps we can rearrange it. First we must consider the error term. We have to assume that it is multiplicative rather than additive. Thus the regression model becomes

$$P_t = P_0(1 + i)^t v_t \quad \text{where } v_t = \text{the error term}$$

To transform this model we take logarithms. Doing this produces

$$\log_{10} P_t = \log_{10} P_0 + t \log_{10}(1 + i) + \log_{10} v_t$$

This may not look very much like our usual linear regression model but if we replace $\log_{10} P_t$ by Y_t; $\log_{10} P_0$ by α because it will be constant; $\log_{10}(1 - i)$ by β because this will also be constant; and $\log_{10} v_t$ by e_t we obtain

$$Y_t = \alpha + \beta t + e_t$$

This is our normal regression equation. We estimate β and taking the antilog of β we obtain $(1 + i)$. It is an easy step to obtain i from this.

For example, with data collected from the years 1876 to 1970 ($n = 93$) the following estimating line was obtained:

$$\log_{10} (\text{expenditure}) = 13.71 + 0.05248t$$

From this we know that $\log_{10} (1 + i) = 0.05248$, and in taking the antilog of this quantity we obtain the relationship $(1 + i) = 1.054$. We find that $i = 0.054$, or the growth factor equals 5.4 percent.

Plotting the raw data, we obtained the plot on the left-hand side of diagram 9.12. Notice that on fitting a linear relationship to the raw data we only managed an R^2 value of 0.55; however, after we performed the log transformation, the R^2 value jumped to an impressive 0.96. We can leave the discussion concerning the error term until the next section.

Diagram 9.12

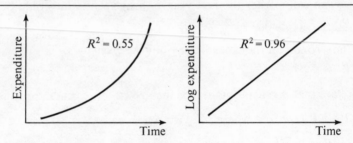

Another common model where the parameters are nonlinear is

$$Y_t = \alpha e^{\beta X_t} v_t \quad \text{where } v_t = \text{the error term}$$

This is one of exponential growth. It may be the case that incidence of terrorism is exhibiting such behavior. The scattergram may look like that shown in diagram 9.13. The fitting of a straight line to this model would

Diagram 9.13

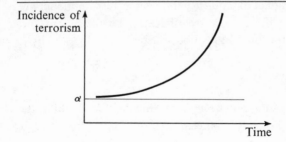

not satisfy, resulting almost certainly in very low values of R^2. However, if we transform the model using natural logarithms we can produce something more serviceable from the point of view of our regression procedure. This transformation would produce

$$\ln Y_t = \ln \alpha + \beta X_t + \ln v_t$$

This is linear and the parameters can be estimated.

These are two examples of models which are intrinsically linear but have nonlinear parameters. There are, of course, many other models and many other ways of transforming them to obtain a linear relationship to exploit our linear techniques.

In summary, it can be argued that most intrinsically linear models of sort discussed in this section can be rendered linear by some form of logarithmic transformation.

9.2.3   The story of the error term

We have been stressing the importance of the error term in regression. This importance should be borne in mind when carrying out the transformations described in the previous two sections. After all transformations, our error term must enter the equation additively. Thus the condition of the error term before the transformation should be such that the transformation makes the error term additive. For instance, look at the previous model

$$Y_t = \alpha e^{\beta X_t} v_t \quad \text{where } v_t = \text{the error term}$$

On transformation, using natural logarithms, this became

$$\ln Y_t = \ln \alpha + \beta X_t + \ln v_t$$

Notice that after the transformation the error term has entered the equation additively. Thus, before the transformation, the error term had to enter the equation multiplicatively. It is important that this always happens. Although it looks very similar to the previous one, the model below has no linear transform

$$Y_t = \alpha e^{\beta X_t} + v_t$$

It is not intrinsically linear. The researcher must seriously consider the effect of transformations upon the error term. Lack of such considerations can produce models violating the assumptions of classical linear regression, with all the consequent difficulties. In essence I am suggesting that the student be constantly aware of the history of the error term. The researcher has to make assumptions about the error term in order to continue. The accuracy of the resulting estimates will reflect the wisdom of the assumptions.

9.3 Restricted coefficients

There are some instances, unfortunately few and far between, where we have prior information concerning the regression coefficients. For example, we may know that the intercept coefficient is zero and that the line goes through the origin. Or we may have determined theoretically that some regression slope is equal to unity. This information can be used with the regression process to aid us in our estimating procedures.

But first we should discuss where such prior information could come from. Basically there are two sources. The first is theory. For example, in the simplest form of an arms-race equation where the arms level of one nation is dependent upon the arms level of another nation, we might initially propose that when the arms level of the one nation is zero, the arms level of the other will also be zero—a model that pacifists have seized upon with confounding regularity! In this situation we see that the line will pass through the origin. Having decided this, we can use the information in our regression.

Another source for obtaining prior information is data. There are occasions when we have determined the relationship between some variable and another in one set of circumstances. We may also be fairly sure about the magnitude of the relationship. We can then use this *a priori* information in any other equation where the two variables are included.

In general, we can say that prior information helps with our inference testing. This accords with common sense. The more we know about a research situation the less we have to depend on the data. And this is reflected in reduced estimates of the residual variance. The student will remember our estimate of the residual variance is given by

$$\hat{\sigma}^2 = \frac{\mathbf{e}'\mathbf{e}}{n - K}$$

Since we know one of the estimates, it follows that the number of estimates that we require, K, is reduced by one. Thus the denominator in this equation is increased, that is, the number of degrees of freedom is increased, reducing the estimate $\hat{\sigma}^2$. Since this estimate is used in obtaining our variance-covariance matrix, we can also conclude that it will have a similar effect on the variances of our regression estimates. Let us look at a few common situations.

In the first situation we restrict the intercept to a fixed value of zero. Thus in the following model the regression line goes through the origin.

$$Y_i = \beta_1 X_{i1} + \beta_2 X_{i2} + u_i$$

Mostly this provides little difficulty for the researcher because many regression program packages have a facility for carrying out this maneuver.

But we can use this example to demonstrate the effect of doing this on the coefficient of determination R^2. First, we know that the normal least-squares regression process will produce the maximum R^2 for any set of data. So any deviation from this process can only reduce the R^2 value. But more than this, as diagram 9.14 demonstrates, it is possible to obtain what at first might seem to be ridiculous, a negative R^2 value. The student will remember from equation (7.17) that R^2 is given by

$$R^2 = 1 - \frac{\sum (Y_i - \hat{Y}_i)^2}{\sum (Y_i - \bar{Y})^2}$$

In the diagram we see that $\sum (Y_i - \hat{Y}_i)^2$ is likely to be larger than $\sum (Y_i - \bar{Y})^2$, and so R^2 can be negative.

Diagram 9.14

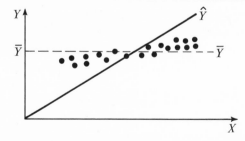

The student should remember that whenever we introduce prior information into the regression process, we cannot rely upon the R^2 value. It becomes meaningless. And this is the case whatever restriction we incorporate.

In another situation we may very well want to restrict the value of a regression slope. This again produces no particular problems. Let us suppose that in the following unrestricted regression model we want to restrict the value of β_2:

$$Y_i = \beta_0 + \beta_1 X_{i1} + \beta_2 X_{i2} + u_i$$

To process this restriction, we arrange the equation so that the restricted segment of the equation is on the left-hand side.

$$(Y_i - \beta_2 X_{i2}) = \beta_0 + \beta_1 X_{i1} + u_i$$

We can then process the following model:

$$Y_i^* = \beta_0 + \beta_1 X_{i1} + u_i$$

where $Y_i^* = (Y_i - \beta_2 X_{i2})$. The regression coefficients obtained have the desirable properties we discussed earlier, and their variance is also reduced. The R^2 value, however, should be treated with the utmost caution.

Thus we see that restrictions on parameters can be accommodated quite easily as long as we appreciate the adverse effects on the coefficient of determination.

9.4　Lagged variables

In many synchronic analyses, models contain variables which are lagged, i.e., are from a previous time point. In these situations either the explanatory variables are lagged or the explained variable is lagged as an explanatory variable. For example, the Stanford Studies for Conflict and Integration collected data for the 38 days of the crisis that led to World War I. These data were on two variables, perceptions of hostility and expressions of hostility. Among the many plausible models, two major ones were

$$P_t = f(E_t, E_{t-1}, E_{t-2}, \ldots, u_t) \quad \text{memory model}$$

$$P_t = f(P_{t-1}, P_{t-2}, \ldots, u_t) \quad \text{habit-persistence model}$$

where P_t is the perception of hostility on day t
　　　E_t is the expression of hostility on day t
　　　u_t is an error term

In the first model, the hypothesis is that perceptions of hostility on any day are a function of expressions of hostility received on that day t plus expressions of hostility on all the previous days of the crisis. This model is called a distributed lag model since the explanation of perceptions of hostility is distributed over the lagged values of expressions of hostility.

In the second model, perceptions of hostility are hypothesized as caused by previous perceptions of hostility. This is a serial correlation model. The explanatory variables are the explained variable lagged.

We shall consider both of these types of model.

9.4.1 Distributed lag models

Let us consider the distributed lag model shown above.

$$P_t = \beta_0 + \beta_1 E_t + \beta_2 E_{t-1} + \beta_3 E_{t-2} + \cdots + u_t$$

A model as comprehensive as this one poses some difficulties. First, there is the question of how far back one should go in including terms in the regression model. If the researcher goes too far back, problems with degrees of freedom will develop. As one increases the number of variables in the equation (K), the quantity $n - K$ will decrease. As a result of this our estimate of the variance of the residuals will become unstable.

To emerge from this problem the researcher can do one of two things. The first is to decide, for theoretical reasons, how far back to go with the lagged variables. For instance, Zinnes[4] proposed that only four of the past days' expressions should be included as explanatory variables. She argued in terms of a memory decay model and suggested that decision makers in the 1914 crisis were unlikely to remember expression of hostility for longer than four days.

Another method of determining a cut-off point for lagged variables is to include only those variables whose regression coefficient is significant at some prescribed level. In this procedure one carries out a preliminary regression with all of the lagged variables. Having selected some significance level, one then excludes from the model all those lagged values which are not significant. This method is exclusively inductive, generating theory by data. It should be used with caution.

Even if one can provide some convincing argument as to how far back one can go with lagged values, there are still other problems. Time-series data almost always suffer with autocorrelation. That is,

$$X_t = f(X_{t-1}, X_{t-2}, \ldots, X_{t-n} + u_t)$$

4. D. Zinnes, "The Expression and Perception of Hostility in Prewar Crisis: 1914," in *Quantitative International Politics*, ed. J. D. Singer (New York: Free Press, 1968), pp. 85–119.

This implies that the explanatory variables in the distributed lag models are correlated, producing all kinds of problems with multicollinearity, not the least of which is the explosion of the variance-covariance matrix. Such problems are discussed in Chapter 7.

In summary, unless the researcher can argue a plausible case for omitting certain lagged variables and then show that there is no autocorrelation between those explanatory lagged variables included, the estimates must be treated with caution. Of course, this is the most conservative position possible. There will be no situation where the researcher does not have to balance accuracy against reality. Distributed lag models are theoretically plausible and will generate the problems referred to above. The researcher should at all times be aware of the difficulties and the kinds of assumptions he has had to make in order to proceed. This awareness is sensible in a politometrician.

9.4.2   Serial correlation

The habit-persistence model,

$$P_t = \beta_0 + \beta_1 P_{t-1} + \beta_2 P_{t-2} + \cdots + u_t$$

has lagged values of the explained variable as explanatory variables. This type of model presents problems similar to the distributed lag models plus some of its own.

First, it becomes difficult to know when to terminate the effects of the lagged variables. As in the previous case, one can either choose some theoretically plausible cut-off point, or one can allow the data to select the level of truncation. Second, since the model itself posits dependence between various values of perceptions of hostility, there will be considerable multicollinearity between explanatory variables. The whole situation is a kind of statistical *Catch 22*. If there is no relationship between consecutive lagged values of the variable then the model is irrelevant, but there is no multicollinearity and the estimates are unbiased. However, if the model is relevant in explaining the perception of hostility on any day, then there will be significant multicollinearity and the estimates thus obtained have inflated variances, making the necessary inferences less dependable. For sure, it is never as bad as this because we can accommodate reasonable levels of multicollinearity.

These difficulties are common to all lag models but serial correlation models have significant problems peculiar to them. The first concerns the error term. Let us explain this using the following very truncated model:

$$P_t = \beta_0 + \beta_1 P_{t-1} + u_t$$

As we have seen, with time-series data it is usually the case that the error terms are related in some way. We saw in Chapter 5 how it is possible to get out of this situation. But when one has a model like the habit-persistence model there are added complications. Given the equation above, we can also put down the relationship

$$P_{t-1} = \beta_0 + \beta_1 P_{t-2} + u_{t-1}$$

Thus we see that P_{t-1} is dependent upon the error term u_{t-1}. And going back to the previous equation, if u_{t-1} is related to u_t, then P_{t-1} will be related to u_t. That is, an explanatory variable will be related to the error term. This invariably produces biased estimates of the regression coefficients. Again this problem is not insurmountable, as we shall see when we consider simultaneous regression equations. Nevertheless, it should be constantly in the mind of the researcher applying regression to serial correlation models.

Finally, one of the assumptions of the regression process is that the consecutive explained variable values should not be related. Obviously this assumption is violated. The consequence of this violation is that the R^2 value will be artificially high.

Again, I have put before the student many of the problems involved with these kinds of models. It sounds pessimistic when listed as above. The student should remember that violation is a matter of degree. The pragmatist will try to avoid too much violation while still attempting to establish some estimates. Economics has these same problems, and econometricians have contrived quite ingenious ways of overcoming them. A student should look to their examples when dealing with a specific model.

9.4.3 Stroboscopic effects in time-series analysis

It is the custom, and indeed a requirement, in time-series analysis that observations of the variables are collected at fixed and regular intervals. Elsewhere, I have articulated my doubts about the effects such customs and requirements have on results. Let me explain my worries in an entirely different context.

Movies are made by photographing movement at 16 frames per second. There is a fixed and regular observation of this movement. When the film is flicked at a certain speed the illusion is one of continuity in the movement of whatever has been photographed. However, consider the situation when there is a particular kind of movement on the screen. More specifically, consider a stagecoach going forward. Because the frequency of the spokes passing the camera is sometimes at odds with the frequency of the frames per second, the wheels of the stagecoach may appear to go forward, be stopped, or indeed go backward. This is because of some stroboscopic effect. Observing the wheels instead of the stagecoach, we might very well make the wrong inference as to its motion. Although it is going forward, we might infer that it is going forward much slower than it really is, that it has stopped, or that it is going backward.

Returning to social processes, is it not conceivable that we might make exactly the same kinds of erroneous inferences, since, when we observe at fixed and regular intervals, we are in essence doing nothing more than making an empirical motion film of our phenomenon? May it not be the case that if we do get such stroboscopic effects then we may get all kinds of positive, negative, and zero relationships, all of which are erronenous and are due to the frequency with which we observe our phenomenon, whether it be by the day, week, month, or year?

What we are, in fact, doing when we observe at such intervals is insert into our process some interval of causation. If the interval of causation matches exactly the true interval of causation, then presumably none of the effects that I have described will occur. But I wonder how often that particular agreeable set of circumstances happens. It is my proposition that underestimation or overestimation of the interval of causation will lead to stroboscopic effects in examining diachronic social processes.

The student might well want to consider this as another cautionary tale for use when dealing with time-series data.

9.5   Stepwise regression procedures

A common question in politometrics is, What is the best regression equation? Of course "best" can have several meanings, but it usually indicates the following two considerations:

(1) the number of explanatory variables in order to obtain a good prediction equation, and
(2) the cost of processing large numbers of explanatory variables.

It is thus a compromise between cost and benefit.

The normal criterion is based upon the amount of variance of the explained variable that is accounted for by the explanatory variables. One only wants to include those explanatory variables that pay for their increase in computation time by substantial increase in variance explained. Thus a variable is included in the model if and only if it contributes a prescribed amount of explained variance. This criterion is obviously completely inductive and as long as the student recognizes it as such there is no need to belabor the point.

The decision to add another variable to the regression equation is made on the basis of the test given in equation (8.6).

$$\frac{(SSR_Q - SSR_R)/\text{no. of variables added}}{SSE_Q/(n - Q)} \overset{d}{=} F \quad \begin{array}{l} \text{no. of} \\ \text{variables added,} \\ n - Q \end{array}$$

Thus variables are added to the equation, and after each addition the equation above is calculated to determine whether the last variable added has increased the amount of variance accounted for over and above a quantity that is statistically significant at some prescribed level. This, in rough terms, is the process that goes on, although the more modern techniques base inclusion or exclusion on the partial correlation of the variable under question with the explained variable, given that the variables already in the equation are kept constant. Specifically there are two stepwise procedures—one called stepwise regression that adds variables, and another called the backwards elimination procedure that includes all explanatory variables and then eliminates them one by one until the required state of grace is reached.

In some of the more sophisticated procedures, variables can enter the regression equation and, after the addition of other variables, leave the equation. A complete description of the various procedures has been provided by Draper and Smith.[5] The student interested in the processes of computation should read their excellent description.

9.6 Missing data and regression

Often the politometrician is confronted with an incomplete data set. This is particularly frustrating when only one of a pair of observations on two variables is missing. For example, we may have an observation for the

5. N. R. Draper and H. Smith, *Applied Regression Analysis* (New York: Wiley, 1967).

explained variable but not for the explanatory variable. The question arises whether to delete the pair of observations altogether and lose some information or whether to try in some way to exploit the partial information.

Most computer regression programs make the decision for the researcher by having subprograms which carry out pairwise deletion. If the researcher opts for this maneuver, it is sensible to know the consequences.

At the outset it should be pointed out that the incomplete information held by one observation of a pair of variables when the other is missing is of little use. One concludes therefore that pairwise deletion is the best strategy. Let us consider two kinds of situations. In the first the explanatory variable is nonstochastic. In this situation, when there are missing observations it can be shown that the efficiency of the estimates is affected. The degree of effect upon the efficiency will reflect the variance of the missing observations. Should this variance be high, so too will be the loss of efficiency; more invariant missing observations have little effect upon efficiency. Losses of efficiency have some detrimental effect upon the hypothesis-testing procedures.

When the explanatory variable is stochastic the same problems exist, although it is possible to use the Monte Carlo walk method for generating the missing data. To do this, the researcher calculates the mean and standard deviation of the available values for the variable with the missing observations. Using one of the many programs provided by computer analysts, the researcher can now generate values coming from a normal distribution with the specified mean and standard deviation. These values are substituted for the missing observations. Using this process, the estimates remain unbiased and the loss of efficiency is reduced.

Further readings

Substantive

Alker, H. A. 1966. "The Long Road to International Relations Theory: Problems of Statistical Nonadditivity." *World Politics* 18: 623.

Broach, G. T. 1972. "Interparty Competition, State Welfare Policies and Nonlinear Regression." *Journal of Politics* 34: 737.

Choucri, N., and R. C. North. 1975. *Nations in Conflict*. San Francisco, Freeman.

Davis, O. T., M. A. H. Dempster, and A. Wildavsky. 1966. "A Theory of the Budgetary Process." *Am. Pol. Sci. Review* 60: 529.

Gray, V. 1973. "Innovation in the States: A Diffusion Study." *Am. Pol. Sci. Review* 67: 1174.

Hollenhurst, J., and G. Ault. 1971. "An Alternative Answer to: Who Pays for Defence?" *Am. Pol. Sci. Review* 65: 760.

Kim, J., and B. C. Koh. 1973. "Electoral Behavior and Social Development in South Korea: An Aggregate Data Analysis of Presidential Elections." *Journal of Politics* 35: 825.

Kort, F. 1968. "A Nonlinear Model for the Analysis of Judicial Decisions." *Am. Pol. Sci. Review* 62: 546.

Kramer, G. H. 1971. "Short-Term Fluctuations in U.S. Voting Behavior, 1896–1964." *Am. Pol. Sci. Review* 65: 131.

Landes, W. L. 1974. "Legality and Reality: Some Evidence in Criminal Procedure." *Journal of Legal Studies* 3: 287.

Paranzino, D. 1972. "A Note on Political Coerciveness and Turmoil." *Law and Society Review* 6: 651.

Seitz, S. T. 1972. "Firearms, Homicides, and Gun Control Effectiveness." *Law and Society Review* 6: 595.

Statistical

Goldberger: chapter 5.
Johnston: chapters 3, 6, and 10; selected portions only.
Kmenta: chapter 11.

There is also one other good piece in time-series data analysis:
Hibbs, D. A. 1974. "Problems of Statistical Estimation and Causal Inference in Times Series Regression Models." In *Sociological Methodology 1973–1974*, ed. H. L. Costner, p. 252. San Francisco: Jossey-Bass.

Part Three

In this final part of the book we consider multivariate, multiequation models. These are simultaneous models mostly and have peculiar problems. We deal with some of these problems, providing the student with rules of thumb by which he can determine the severity of the problems. Some consideration is also given to variations of simultaneous models such as path and causal inference models.

10
multiequation recursive models

In the first part of this book we considered research situations where there were the minimum of variables—one explained and one explanatory variable. In the second part we increased the number of explanatory variables in our models. In this part of the book we shall increase the number of equations in our politometric models. We thus move into the more realistic multiequation, multivariate research mode. We shall see, as we proceed, that integrated sets of equations provide estimation problems peculiar to themselves. It is not just a matter of processing each equation individually. Because variables in one equation may very well appear in another equation in the multiequation model, the equations are not independent. These links between equations deny us the ability to treat each equation individually. We are truly dealing with an integrated politometric model rather than a set of unrelated equations.

Political scientists are, or should be, concerned with causal effects. This concern has accelerated with the development of causal inference techniques during the early 1960s. Techniques put forward by Hubert Blalock and Sewall Wright have been increasingly used by sociologists and political scientists, but not, it should be pointed out, with the same avarice by econometricians. One reason for this is that both techniques assume more than econometricians care to about their models. For instance, except in a few cases, only recursive models are processed. In these models, feedback, neither direct nor indirect, is catered to. Many econometricians and an increasing number of other social scientists consider such models unrealistic, and we will deal with more realistic models later. Nevertheless, there are some situations where these models are valid, and in the case of the path-analysis technique developed by Wright, many of the parameters developed have a more intuitive meaning to the researcher. In this section we will discuss both the method put forward by Blalock[1] and that developed by Wright.[2] But first a discussion of recursive models.

10.1   Recursive models

A recursive model is one in which causal effects are hierarchal. In such models the direction of influence from any particular variable is not allowed to feed back to that variable either directly or indirectly. Consider the causal network shown in diagram 10.1. In this network there are four variables. Notice that A is more causally prior to B, which is more causally prior to C, which is more causally prior to D. And since cause is denoted by the direction of the arrows, we see that there is no way that we can arrive back at any variable after leaving it without going against one of the arrows.

Diagram 10.1

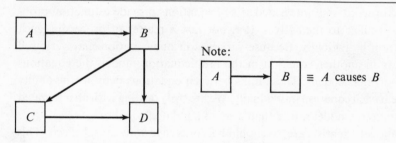

This recursive characteristic of these techniques has led to much controversy mainly among those who feel that there are no social systems in which either direct or indirect feedback does not take place. The argument proceeds, What is the use of the causal inference techniques? Further, since other econometric techniques, as we shall see later, can

1. H. M. Blalock, *Causal Inference Analysis.*

2. S. Wright, "The Method of Path Coefficients," *Annals of Mathematical Statistics* 5 (1934): 161–215.

cope with feedback, why even bother with these recursive techniques? The proponents of the techniques put forward the following types of arguments. First, the techniques bridge the gap between verbal theories and research techniques. Second, as we shall see, ordinary least-squares are applicable, and so these techniques become available to the less mathematically inclined researcher. Third, they provide some rules for making causal inferences on the basis of empirical relationships.

It has also been argued that in reality only recursive models exist. Since we deal with causal relationships which have as their implicit foundation some "if . . . then" statement, there is an implied passage of time, no matter how small. Philosophers argue that cause cannot act at a distance. There is a continuous chain of action and reaction along some causal chain or sequence; at a micro level, cause and effect will be adjacent. Normally, however, we cannot measure at these tiny levels and thus the micro causal processes are invisible to us. We measure at more aggregated levels. For instance, if the sun went out at this minute, it would take eight minutes before we knew of it on earth. Obviously, there has been a complete causal chain at the smallest level in which cause and effect has been adjacent, that is, as the energy or lack of energy was transmitted from atom to atom between the sun and the earth. This is consistent with the philosopher's point. But we cannot observe all of these and may very well only be concerned with the behavior of the sun and behavior of the earth. Thus our causal sequence would be an eight-minute gap.

Let us take this a little further. Suppose we have the following sequence. Attitudes at time t go through some causal sequence and produce a behavior at time $t + 1$ which, in turn, goes through some other causal sequence to produce an attitude at time $t + 2$, and so on. This can be shown by diagram 10.2. Notice that we have not violated the philosopher's

Diagram 10.2

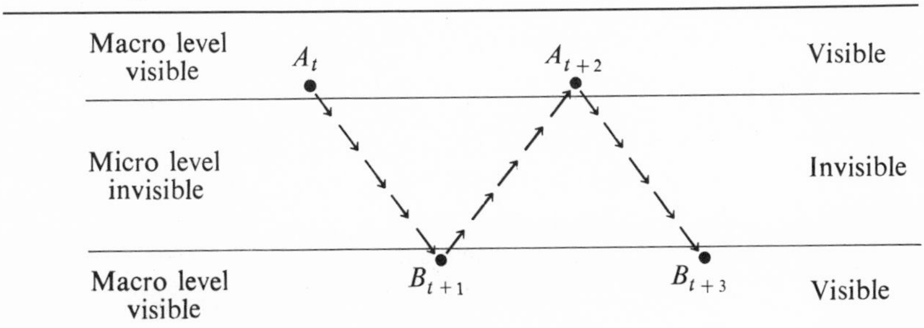

insistence upon cause acting adjacently. We allow this, but we also notice that all of the minute causal processes are invisible to us. The variables A_t, B_{t+1}, and A_{t+2}, and so on are visible to us. This is a recursive model, and indeed, it is difficult to imagine how there can be anything other than recursive models, since we will never be able to make visible all of the causal processes that go on at the minutest levels.

Ostensibly this is feedback, but it is feedback at a much later time. If we treat A_t as a different variable from A_{t+2}, etc., and B_{t+1} as a different variable from B_{t+3}, etc., then we have a recursive system. It would seem, then, that the recursivists have the advantage at the moment. But the argument can be taken further.

Our difficulty so far has been the level of the variable that we can accommodate. Thus the variables between A_t and B_{t+1} are not visible to us because we work at a more macro level. However, suppose we now consider the level of data we work at, and suppose further that we cannot measure down to the levels of t, $t+1$, $t+2$, and so on, but only measure at some aggregated time period, say T. Diagram 10.3 shows this possibility for our previous example. Notice that in the time period T for which we have data, there have been several interactions between A and B. And since we have aggregated all of these causal paths by measuring at a frequency far less than that of the processes going on, we will infer the model shown in diagram 10.4. That is, A and B are causing each other simultaneously. Now we know that this is not true but an artifact of the measurement of the variables. Had we taken observations at a time frequency of t we would have caught the true recursive model. It is the missing of the true time frequency that has forced us to consider the nonrecursive model.

Diagram 10.3

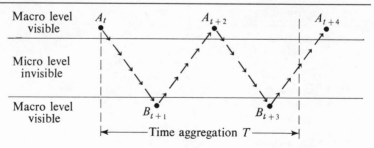

| Macro level visible | A_t | | A_{t+2} | | A_{t+4} |

| Micro level invisible | | | | | |

| Macro level visible | | B_{t+1} | | B_{t+3} | |

←——— Time aggregation T ———→

Diagram 10.4

Whether we have a recursive or nonrecursive model depends entirely upon the time aggregation with which we operate. If we have smaller periods of measurement, we are more likely to get the true model which will always be recursive. As we move from this ideal set of circumstances toward one where we have large time aggregation and finally cross-sectional analyses, so we move away from the true recursive model and toward the nonrecursive model. All of this assumes that our level of variable is macro rather than micro.

So whether to use causal inference techniques with recursive models and politometric techniques with nonrecursive models hinges upon the time aggregation of the data. Admittedly it is our inability to collect data at the correct time interval of causation which requires us to develop the more technical methods required for coping with nonrecursive models.

One more consideration in this argument concerns systems that have reached a position of equilibrium. At this point, consecutive values of the same variable are the same. Thus in our example

$$A_t = A_{t+2} = \cdots = A_T \quad \text{and} \quad B_{t+1} = B_{t+3} = \cdots = B_T$$

Such a phenomenon would give the appearance of simultaneous causation between the two variables, and the model with simultaneous feedback would describe the situation adequately. The example of three steel balls in a bowl is always given as an example of simultaneous causation. In this situation the length of the time interval of measurement would have little effect on the results as long as the system was in equilibrium for the whole of that time interval. Nevertheless, such systems have to attain this equilibrium and they are sometimes disturbed away from the equilibrium

point. Should either of these situations occur, then we return to the
recursive process described before.

10.2   Causal inference analysis—Simon-Blalock method

In a seminal work on the correlational relationship between three variables, Simon[3] argued that if we control for Z then we should obtain zero
correlation between variables X and Y in two situations (diagram 10.5).
The situation on the left is called spurious correlation, and that on the
right is called the intervening variable situation. The term "control for"
denotes a statistical process for holding constant a particular variable
while correlating others. The correlation would generally be shown as
$r_{YX \cdot Z}$. In this subscript notation it is usual to show the explained variable
first and the explanatory variable second, followed by a period with the
controlled-for variables after it. Blalock[4] extended this argument to include more than three variables and, for any causal network, he put
forward a procedure which allowed the prediction of zero partial correlations. The goodness of the model depended upon the accuracy of these
predictions when the partial correlations were calculated from the data.

Diagram 10.5

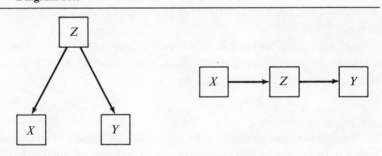

Consider the recursive model shown in diagram 10.6. Notice that it has
no feedback, neither direct nor indirect. Also notice that all possible
arrows are present. The sets of equations that represent these relationships

3. H. A. Simon, *Models of Man* (New York: Wiley, 1957).

4. H. M. Blalock, *Causal Inference Analysis.*

Diagram 10.6

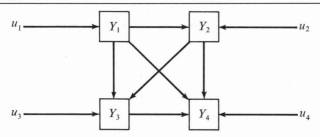

are called structural sets. The structural set for this model is

$$Y_{i1} = \beta_{10}Y_{i0} + \qquad\qquad\qquad\qquad\qquad u_{i1}$$
$$Y_{i2} = \beta_{20}Y_{i0} + \beta_{21}Y_{i1} + \qquad\qquad\qquad u_{i2}$$
$$Y_{i3} = \beta_{30}Y_{i0} + \beta_{31}Y_{i1} + \beta_{32}Y_{i2} + \qquad u_{i3}$$
$$Y_{i4} = \beta_{40}Y_{i0} + \beta_{41}Y_{i1} + \beta_{42}Y_{i2} + \beta_{43}Y_{i3} + u_{i4} \qquad (10.1)$$

In these equations the first subscript of the regression slopes denotes the explained variable, the second indicates the explained variable to which it is attached. All other notation is consistent with previous use. Notice how in these equations we have a causal hierarchy. Each variable enters the system first as an explained variable and then goes over to the other side of the next equation to become an explanatory variable. The student should remember that Y_{i0} is the constant term.

The Blalock method forces us to place one restriction on the structural set. The restriction we make concerns the error terms. We do not allow the error-term distributions to be correlated. Thus the distribution u_{i1} is independent of the other error distributions, u_{i2}, u_{i3}, and u_{i4}. The reason for this derives from our prohibition on explanatory variables in any one equation being related to the error term in that same equation. If u_{i1} is related to u_{i3} and Y_{i1} is dependent upon and thus related to u_{i1}, Y_{i1} will also be related to u_{i3}. Thus in the third equation we have an explanatory variable related to the error term. This is not allowed.

Unfortunately in this fully recursive model all of the causal arrows are present and we are unable to make any prediction about the regression slope coefficients being zero. This will only occur when causal arrows are missing. Suppose, however, that we thought that the true causal network was as shown in diagram 10.7. Compared with the fully recursive set we

Diagram 10.7

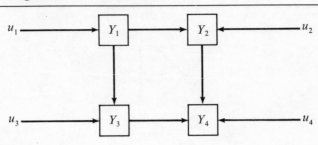

can see that

(1) Y_3 is not related to Y_2,
(2) Y_4 is not related to Y_1.

These two statements provide our two prediction equations in the testing of this model. These prediction equations are

$$\beta_{32} = 0$$
$$\beta_{41} = 0$$

The technique now is to carry out ordinary least-squares on the fully recursive model and determine whether the data confirm the predictions. If they do, then the model is not rejected; if they do not, the model is rejected in the favor of another model.

The student should be fully aware of the procedure here and the steps that are taken. The first step involves the theoretical development of a model and its display in some form of arrow causal network. This is then converted into a mathematical formulation which allows the zero-slope coefficient predictions to be generated. The predictions are then tested by confronting the fully recursive model with the data and carrying out ordinary least-squares. The slope coefficients which are predicted to be zero are then tested for their closeness to zero.

The technique discussed above is basically similar to that suggested by Blalock. There is one fundamental difference. He deals with partial correlation coefficients. In the above example his predictions would be:

(1) $r_{Y_3 Y_2 \cdot Y_1} = 0$
(2) $r_{Y_4 Y_1 \cdot Y_2 Y_3} = 0$

However, if these predictions are confirmed then the corresponding regression-type predictions will also be confirmed. Since we have emphasized regression estimation as against correlation estimation, we shall continue with this emphasis here. The results will be the same.

There is a problem of discrimination between models. Let us return to the three-variable case to discuss this point. Suppose we are trying to discriminate between the two models in diagram 10.8. The regression set for the spurious model would be

$$Y_{i1} = \beta_{10} Y_{i0} + \qquad\qquad\qquad u_{i1}$$
$$Y_{i2} = \beta_{20} Y_{i0} + \beta_{21} Y_{i1} + \qquad\qquad u_{i2}$$
$$Y_{i3} = \beta_{30} Y_{i0} + \beta_{31} Y_{i1} + 0 \cdot Y_{i2} + u_{i3}$$

The prediction here is

$$\beta_{32} = 0$$

The regression set for the intervening model is

$$Y_{i2} = \beta_{20} Y_{i0} + \qquad\qquad\qquad u_{i2}$$
$$Y_{i1} = \beta_{10} Y_{i0} + \beta_{12} Y_{i2} + \qquad\qquad u_{i1}$$
$$Y_{i3} = \beta_{30} Y_{i0} + 0 \cdot Y_{i2} + \beta_{31} Y_{i1} + u_{i3}$$

The prediction here is

$$\beta_{32} = 0$$

This is the same prediction as in the spurious model. The technique has no way of discriminating between the two models. Only additional information can help us in this situation. This problem is multiplied as the number of variables in the model increases, so that if we cannot even discriminate between two models in the three-variable case, we may very well not be able to discriminate between a larger number of models when we increase the number of variables.

Diagram 10.8

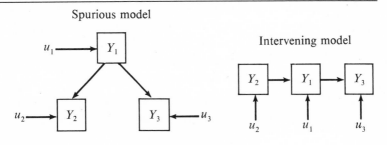

Spurious model

Intervening model

Let us recapitulate. The technique is based upon assumptions that the the system is recursive and that the error terms in each equation are unrelated to one another. Having accepted these assumptions, we are then able to make predictions concerning the regression slopes of any missing links from the fully recursive model. There will be as many predictions as there are missing links from the fully recursive set. Obviously, the more predictions we confront, the more we would rely upon the model.

10.3   Causal inference analysis—path coefficients

Sewall Wright[5] developed the following technique for determining relative influences of variables within any model. The actual measure is called a path coefficient. This coefficient gives the researcher some idea of the magnitude of the influence of one variable on another, and what is more important, can also determine the influence of the variables left out of the model, that is, the error terms. Let us use the model in diagram 10.9 as an example.

Diagram 10.9

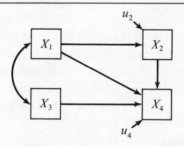

Notice again that this is a recursive model, although techniques have been devised to take care of nonrecursive models. Also notice that we can include correlations between variables if we so desire. These are indicated by curved lines with an arrow at each end. In this way we can say about variables X_1 and X_3 that we do not know which is causing which, and indeed are not really bothered. Both are accommodated by the correlation. The normal regression equations for this would be

5. S. Wright, "Method of Path Coefficients."

$$X_{i2} = \beta_{20}X_{i0} + \beta_{21}X_{i1} + \qquad\qquad\qquad u_{i2}$$
$$X_{i4} = \beta_{40}X_{i0} + \beta_{41}X_{i1} + \beta_{42}X_{i2} + \beta_{43}X_{i3} + u_{i4}$$

Suppose, however, that instead of operating with the raw data we standardize the data on each variable. This is done by subtracting the mean of a distribution from a particular observation in that distribution and dividing this quantity by the standard deviation of the distribution. This is carried out for all of the variables. For the sake of this argument we will dispense with the subscript i. The regression equations now become, using Y_1 to represent the standardized X_1; Y_2 as the standardized X_2, and so on:

$$Y_2 = p_{21}Y_1 + p_{2u_2}u_2 \qquad\qquad\qquad (10.2)$$

$$Y_4 = p_{41}Y_1 + p_{42}Y_2 + p_{43}Y_3 + p_{4u_4}u_4 \qquad\qquad (10.3)$$

The diagram thus changes to diagram 10.10. Notice that the intercept term has disappeared—this will always happen when we carry out regression on standardized data. Also notice that we now use p as the regression coefficient; p is the path coefficient and is really the regression coefficient for the standardized data. Finally, notice that we have also allowed a path coefficient for the error terms u_2 and u_4. We cannot obtain these by regression, but we can in other ways, as we shall now demonstrate.

Diagram 10.10

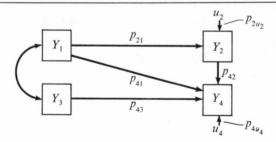

In Chapter 2 we defined a correlation between X and Y using the relationship

$$r_{XY} = \frac{\frac{1}{n}\sum(X_i - \bar{X})(Y_i - \bar{Y})}{\sigma_X\sigma_Y}$$

This can be taken also as

$$r_{XY} = (\text{standardized } X)(\text{standardized } Y)$$

Let us now return to our regression equations of the standardized form. We will take equation (10.2) first.

If we multiply both sides of this equation by Y_1, we will produce the following equation:

$$Y_2 Y_1 = p_{21} Y_1 Y_1 + p_{2u_2} u_2 Y_1$$

But we know that $Y_2 Y_1 = r_{21}$ because Y_2 and Y_1 are standardized. Similarly, $Y_1 Y_1 = r_{11} = 1$. From the diagram we see that u_2 and Y_1 are uncorrelated, so $r_{u_2} Y_1 = 0$. The equation reduces to

$$r_{21} = p_{21}$$

Let us take equation (10.2) again and multiply both sides by Y_2.

$$Y_2 Y_2 = p_{21} Y_1 Y_2 + p_{2u_2} u_2 Y_2$$

But we know that

$$Y_2 Y_2 = 1 \quad Y_1 Y_2 = r_{21} = p_{21} \quad u_2 Y_2 = p_{2u_2}$$

So we obtain another equation in terms of the path coefficients,

$$1 = p_{21}^2 + p_{2u_2}^2$$

Let us now look at equation (10.3). We will multiply this equation by each of the variables on the right-hand side of the equation excluding the error term. This produces

$$Y_4 Y_1 = p_{41} Y_1 Y_1 + p_{42} Y_2 Y_1 + p_{43} Y_3 Y_1 + p_{4u_4} u_4 Y_1$$
$$Y_4 Y_2 = p_{41} Y_1 Y_2 + p_{42} Y_2 Y_2 + p_{43} Y_3 Y_2 + p_{4u_4} u_4 Y_2$$
$$Y_4 Y_3 = p_{41} Y_1 Y_3 + p_{42} Y_2 Y_3 + p_{43} Y_3 Y_3 + p_{4u_4} u_4 Y_3$$
$$Y_4 Y_4 = p_{41} Y_1 Y_4 + p_{42} Y_2 Y_4 + p_{43} Y_3 Y_4 + p_{4u_4} u_4 Y_4$$

and since we know that

$Y_4 Y_1 = r_{41}$	$Y_1 Y_1 = r_{11} = 1$	$Y_2 Y_1 = r_{21}$
$Y_4 Y_2 = r_{42}$	$Y_1 Y_2 = r_{12}$	$Y_2 Y_2 = r_{22} = 1$
$Y_4 Y_3 = r_{43}$	$Y_1 Y_3 = r_{13}$	$Y_2 Y_3 = r_{23}$
$Y_4 Y_4 = r_{44} = 1$	$Y_1 Y_4 = r_{14}$	$Y_2 Y_4 = r_{24}$

$$Y_3 Y_1 = r_{31} \qquad u_4 Y_1 = 0$$

$$Y_3 Y_2 = r_{32} \qquad u_4 Y_2 = 0$$

$$Y_3 Y_3 = r_{33} = 1 \qquad u_4 Y_3 = 0$$

$$Y_3 Y_4 = r_{34} \qquad u_4 Y_4 = p_4 u_4$$

we arrive at the four further equations in path coefficients and simple correlations.

$$r_{41} = p_{41} \qquad + p_{42} r_{21} + p_{43} r_{31}$$

$$r_{42} = p_{41} r_{12} \quad + p_{42} \qquad + p_{43} r_{32}$$

$$r_{43} = p_{41} r_{13} \quad + p_{42} r_{23} + p_{43}$$

$$r_{44} = 1 = p_{41}^2 + p_{42}^2 \qquad + p_{43}^2 \qquad + p_{4u_4}^2$$

We thus have six equations and six unknowns. The unknowns are the path coefficients and can be solved algebraically from the six equations. All we need to do is compute the correlation coefficients between the variables in the model and use these algebraically to produce the path coefficients.

The above method for obtaining the path coefficients is rather tedious. They can be obtained far more quickly by the following rule. To obtain the path equation for any correlation r_{ij} read back from variable i and then read forward to variable j. While doing this, form the products of the path coefficients along the path. Do this for all possible paths from variable i to variable j, going backward from i then forward to j. In any particular path traverse, do not go through the same variable twice. The correlation curves between two variables can be treated as going both ways.

This seems very confusing and we can clarify it using our example. We shall take each traverse in turn to see how the equation is formed. Let us consider the correlation r_{21} first. There is only one way back from Y_2 to Y_1 and this is along p_{21}. Thus we get

$$r_{21} = p_{21}$$

Let us take r_{42}. There are three paths from Y_4 to Y_2. These are straight back from Y_4 to Y_2, backward from Y_4 to Y_1 and then forward to Y_2, backward from Y_4 to Y_3 to Y_1 and then forward to Y_2. Traversing all of these paths and summing, we obtain

$$r_{42} = p_{41} r_{12} + p_{42} + p_{43} r_{32}$$

We can now take r_{41}. There are three paths linking these two variables. The first is straight from Y_4 to Y_1, the second is backward from Y_4 to Y_3 and then to Y_1. The third path is backward from Y_4 to Y_2 to Y_1. Traversing all of these paths and accumulating the path products, we obtain

$$r_{41} = p_{41} + p_{42}r_{21} + p_{43}r_{31}$$

Let us take r_{43}. There are three paths between Y_4 and Y_3. These are backward straight from Y_4 to Y_3, backward through Y_2 and Y_1 and then to Y_3, and finally backward from Y_4 to Y_1 and then to Y_3. Traversing all these paths and accumulating produces the equation

$$r_{43} = p_{41}r_{13} + p_{42}r_{23} + p_{43}$$

So far in this process we have four equations. The final two are obtained by taking each of the two explained variables in turn and squaring every path that goes into it and letting the sum equal one. Thus for Y_2 we get

$$p_{21}^2 + p_{2u_2}^2 = 1$$

Similarly, for Y_4 we get

$$p_{41}^2 + p_{42}^2 + p_{43}^2 + p_{4u_4}^2 = 1$$

We now have six equations using this method, exactly the same number as by the previous method. Comparison of the two sets of equations will show they are similar; the only difference is that in some of the equations r_{23} is replaced by $p_{21}r_{31}$. These are obviously similar amounts and so the two techniques for getting the equations produce the same results.

The path coefficient is rather like a correlation in quality. Whereas the normal regression coefficient is tied to the units in which the data are measured, the path coefficients, since they are standardized regression coefficients, do not have any scale. They are comparable within a model, showing the relative causal importance of various explanatory variables which cause an explained variable. We can gauge which explanatory variable has the greatest influence upon a particular explained variable.

Secondly, the technique allows us to get a similar measure of the importance of the variables specifically left out of the model and included in the error term. This is not as vividly done in any of the other techniques we have seen.

There are other gains to the technique, particularly in situations where we cannot measure all of the variables. Take the situation shown in

Diagram 10.11

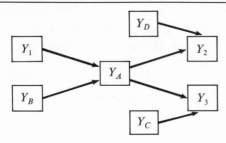

diagram 10.11. In this model the variables that we cannot measure are denoted by alphanumeric suffixes. Notice that we can measure less than half of the seven variables in the model. We can consider variables Y_B, Y_C, and Y_D as variables left out of the system; in essence they are error terms. The variable Y_A is an intervening variable between variables Y_1 and Y_B and variables Y_2 and Y_3. Notice that there are six unknown paths, and for a unique solution we require six equations. The only equations possible are

$$r_{32} = p_{3A}p_{2A} \quad 1 = p_{2D}^2 + p_{2A}^2$$

$$r_{21} = p_{2A}p_{A1} \quad 1 = p_{3C}^2 + p_{3A}^2$$

$$r_{31} = p_{3A}p_{A1} \quad 1 = p_{A1}^2 + p_{AB}^2$$

Notice that we have six of them and therefore we can easily solve algebraically for the six unknown path coefficients.

In another quite fundamental way the technique of path analysis differs from Blalock's causal inference analysis. In path analysis we can allow for correlation between error terms. Consider the model in diagram 10.12. In this model Y_A and Y_B can be considered as variables left out of the

Diagram 10.12

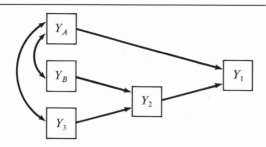

system, or error terms. Notice that there is a two-headed arrow between them, denoting that they are correlated. We can also see that there is a correlational relationship between Y_A and the exogenous variable Y_3; this was not allowed in our previous models. In this model we have six unknowns; four of them are path coefficients and two are correlations. We obviously need six equations to obtain unique solutions for these unknowns. The six equations below provide enough information to allow the solution.

$$r_{12} = p_{12} + p_{1A}r_{AB}p_{2B} + p_{1A}r_{A3}p_{23} \quad r_{11} = 1 = p_{1A}^2 + p_{12}^2$$

$$r_{13} = p_{21}p_{32} + p_{1A}r_{3A} \qquad\qquad r_{22} = 1 = p_{2B}^2 + p_{23}^2$$

$$r_{23} = p_{23} \qquad\qquad\qquad\qquad r_{2A} = 0 = p_{23}r_{3A} + p_{2B}r_{BA}$$

This example provides another rule for developing path equations. One cannot trace a path from one variable to another going through two correlation paths. In this example we cannot traverse from Y_1 to Y_3 by going through the path Y_2 to Y_B to Y_A to Y_3. Nor can we traverse from Y_2 to Y_3 through the path Y_B to Y_A to Y_3.

We see then that path analysis allows theoretical formulations not normally allowed by other techniques. There is, however, one severe difficulty. This has to do with the number of equations compared to the number of unknowns in the model. For a unique solution of the unknowns we must have exactly the same number of unknowns as we have equations. If we have more unknowns than equations, solutions for the unknowns are not available. If we have more equations than unknowns, then we have more than one solution for some of the unknowns. The problem is how to distinguish between the different values of the unknowns.

If we have more equations than unknowns we can quite easily fabricate, because the technique requires it, another unknown. In such a situation the technique is determining the theory—a very dangerous maneuver. Let us look at the example in diagram 10.13, which is similar to the previous example except that the correlations between the error terms and the exogenous variables are omitted. In this situation we have four unknowns and, as we see below, five equations.

$$r_{12} = p_{12} \qquad r_{11} = 1 = p_{12}^2 + p_{1A}^2$$

$$r_{23} = p_{23} \qquad r_{22} = 1 = p_{23}^2 + p_{2B}^2$$

$$r_{13} = p_{12}p_{23}$$

Diagram 10.13

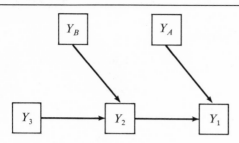

This means that two of the paths can be determined in different ways. That is,

$$p_{23} = r_{23} \quad \text{or} \quad p_{23} = r_{13}/p_{12}$$

and

$$p_{12} = r_{12} \quad \text{or} \quad p_{12} = r_{13}/p_{23}$$

This is an intolerable situation from which one can extricate oneself only by the fabrication of correlations between Y_A and Y_3 and Y_A and Y_B, whether or not such fabrication is theoretically justified.

Finally, path analysis can be used to "test" models in the same sense as the Blalock method. Take the example in diagram 10.14. The path equations are

$$r_{43} = p_{43}$$
$$r_{32} = p_{32} + p_{31}r_{12}$$
$$r_{31} = p_{31} + p_{32}r_{12}$$
$$r_{33} = 1 = p_{31}^2 + p_{32}^2 + p_{3u_3}^2$$
$$r_{44} = 1 = p_{43}^2 + p_{4u_4}^2$$

Diagram 10.14

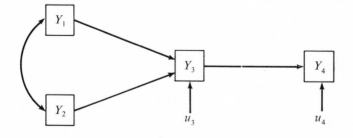

We have five unknowns and five equations. Suppose that from our data we obtain the following correlations:

$$r_{14} = 0.60 \qquad r_{12} = 0.80 \qquad r_{13} = 0.82$$
$$r_{24} = 0.48 \qquad r_{34} = 0.60 \qquad r_{23} = 0.80$$

Solving, we get

$$p_{43} = 0.60; \quad p_{32} = 0.40; \quad p_{31} = 0.50; \quad p_{3u_3} = 0.77; \quad p_{4u_4} = 0.80$$

Notice in this model how the error term u_4 has more leverage on Y_4 than does Y_3 and similarly how the error term u_3 has more leverage on Y_3 than either Y_1 or Y_2 individually: a typical social-science model with more left out of the model than there is in it!

Having developed these path coefficients, we can now proceed to "test" the model. Since there are links missing between Y_4 and the two variables Y_1 and Y_2, we can make the predictions

$$p_{42} = 0 \qquad p_{41} = 0$$

Let us suppose for the minute that these links are included. The model would be as in diagram 10.15. Using this model we produce the extra two equations,

$$r_{42} = p_{42} + p_{32}p_{43} + p_{43}p_{31}r_{12} + p_{41}r_{12}$$
$$r_{41} = p_{41} + p_{43}p_{31} + p_{43}p_{32}r_{12} + p_{42}r_{12}$$

But our predictions are $p_{41} = 0$ and $p_{42} = 0$; thus the equations reduce to two prediction equations,

$$\hat{r}_{42} = p_{32}p_{43} + p_{43}p_{31}r_{12}$$
$$\hat{r}_{41} = p_{43}p_{31} + p_{43}p_{32}r_{12}$$

Diagram 10.15

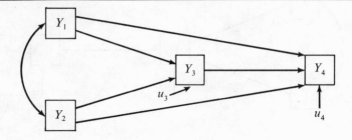

On inserting our already obtained values for the component on the right-hand side of both equations, we get

$$\hat{r}_{42} = (0.40)(0.60) + (0.60)(0.5)(0.80) = 0.24 + 0.24 = 0.48$$

$$\hat{r}_{41} = (0.60)(0.50) + (0.60)(0.40)(0.80) = (0.30) + (0.192) = 0.492$$

But the "true" values of r_{42} and r_{41} are 0.48 and 0.60 respectively. We conclude that the model has failed the test in the case of predicting the value of r_{41}.

The next question is, Can this failure provide us with some clue as to a further, perhaps better model? Obviously, with the paths we have between variable Y_4 and Y_1 we do not account for the "total" covariance between them. If we add a direct path between them we could soak up the remaining covariance. This would produce the model in diagram 10.16.

Diagram 10.16

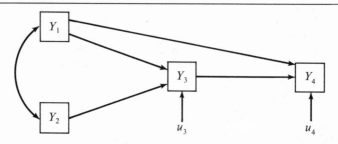

So we see that path analysis can promote theory construction. The danger in this is that we have a one-way bulldozing of theory by data—an inductive process. The sensible scientist will be aware of this danger and proceed accordingly.

Further readings

Substantive

Beyle, T. L. 1965. "Contested Elections and Voter Turnout in a Local Community: A Problem in Spurious Correlation." *Am. Pol. Sci. Review* 59: 111.

Bowers, W. J., and R. G. Salem. 1972. "Severity of Formal Sanctions as a Repressive Response to Deviant Behavior." *Law and Society Review* 6: 427.

Cameron, D. R., J. S. Hendricks, and R. I. Hefferbert. 1972. "Urbanization, Social Structure and Mass Politics: A Comparison within Five Nations." *Comparative Political Studies* 5: 259.

Cnudde, C., and D. J. McCrone. 1966. "The Linkage between Constituency Attitudes and Congressional Voting Behavior: A Causal Model." *Am. Pol. Sci. Review* 60: 66.

Forbes, H. D., and E. R. Tufte. 1968. "A Note of Caution in Causal Modelling." *Am. Pol. Sci. Review* 62: 1258.

Goldberg, A. S. 1966. "Discerning a Causal Pattern among Data on Voting Behavior." *Am. Pol. Sci. Review* 60: 913.

Hilton, G. T. 1972. "Causal Inference Analysis: A Seductive Process." *Administrative Science Quarterly* 17: 44.

Jacobson, A. L. 1973. "Intrasocietal Conflict: A Preliminary Test of a Structural-Level Theory." *Comparative Political Studies* 6: 62.

McCally, S. P. 1966. "The Governor and His Legislative Party." *Am. Pol. Sci. Review* 60: 923.

McCrone, D. J., and C. F. Cnudde. 1967. "Toward a Communications Theory of Democratic Political Development: A Causal Model." *Am. Pol. Sci. Review* 61: 72.

Muller, E. N. 1970. "Cross-National Dimensions of Political Competence." *Am. Pol. Sci. Review* 64: 792.

Seitz, S. T. 1972. "Firearms, Homicides, and Gun Control Effectiveness." *Law and Society Review* 6: 595.

Sharkansky, I. 1968. "Agency Requests, Gubernatorial Support and Budget Success in the State Legislatures." *Am. Pol. Sci. Review* 62: 1220.

Winham, G. R. 1970. "Political Development and Lerner's Theory: Further Test of a Causal Model." *Am. Pol. Sci. Review* 64: 810.

Statistical

Blalock, H. M. 1964. *Causal Inference Analysis in Non-experimental Research.* Chapel Hill, N.C.: Univ. of North Carolina Press.

Blalock, H. M. 1971. *Causal Models in the Social Sciences.* Chicago: Aldine.

Kmenta: pp. 538–39.

Wonnacott and Wonnacott: p. 193.

Wright, S. 1931. "Statistical Methods in Biology." *Journal of the American Statistical Society* 26: 155.

Notice that these econometricians have not written much on recursive models. Politometricians should be intrigued by this fact.

11
structural models and the identification problem

The structural models we have so far discussed are basically unrealistic, for the reason that they do not allow feedback between variables. More realistic models allow such feedback. The aim of this chapter is to provide an example of a political science model which allows feedback. We shall also demonstrate in this chapter how feedback generates problems for our estimation process. The first is the problem of inconsistency in our estimates. Simultaneous equations, which all of these models are, invariably produce equations in which one of the explanatory variables covaries with the error term. The second problem is one of identification. Simply put, we have more unknowns than we have equations, and a consequence of this is an inability to obtain estimates. However, this is not a simple problem and we shall spend much time discussing it. The whole point of this chapter is to set us up for the final chapter in which we produce a complex analysis of the problems and then proceed to provide maneuvers for overcoming them. But first the example.

11.1   An example of a multiequation model

In reality, given any set of covarying variables, there will be times when a particular variable is both an explained and an explanatory variable; there will be times when we will require feedback in the model as against a hierarchical causal network, since some variables are causally prior to others.

As an example, look at a model discussed in Chapter 1. Ted R. Gurr has developed a complex model relating variables concerned with civil strife. Diagram 11.1 indicates the various hypothesized causal connections between the variables. Notice that we have four shapes in this diagram: triangle, rectangle, circle, and diamond. Each of these shapes refers to a type of variable. The triangle refers to an *exogenous* variable. These variables only explain—they are never explained. That is, arrows leave them and never go to them. They are determined outside of the system of the variables. Rectangles in the diagram refer to *endogenous* variables, those variables determined entirely within the system of variables. Notice that these variables can be explained as well as explanatory. That is, arrows go into, as well as away from, these variables. The circle refers to a variable determined without error. Thus social tension is seen as some addition or multiplication of the three variables—stress, strain, and con-

Diagram 11.1

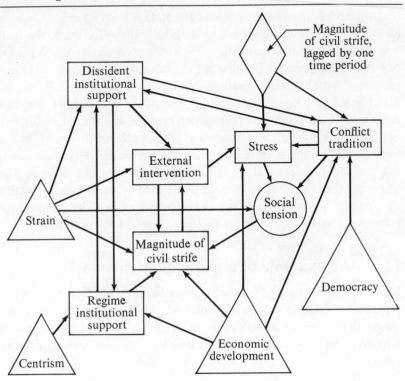

flict tradition—and moreover, social tension is exactly determined by this relationship. Finally, the diamond refers to a lagged variable. The one shown in this diagram is the endogenous variable, magnitude of civil strife lagged by one time period. In essence it acts as an exogenous variable. It has been the convention that all exogenous variables, both lagged and nonlagged, are termed *predetermined*. That is, they are determined outside the system.

Whenever these different types of variables are incorporated in sets of equations, called structural sets, there is a convention for which type of variable gets which symbol. We can put down the equations for the system of equations represented above and identify the symbol representing the types of variables. There will be one equation for each endogenous variable, so we will have seven equations. Each endogenous variable will appear once on the left-hand side of the equations, as an explained variable, but it may very well appear more than once on the right-hand side of the equation as an explanatory variable. In the following system,

X_1 = strain
X_2 = economic development
X_3 = centrism
X_4 = democracy
X_5 = magnitude of civil strife, lagged one time period
Y_1 = regime institutional support
Y_2 = dissident institutional support
Y_3 = conflict tradition
Y_4 = stress
Y_5 = external intervention
Y_6 = magnitude of civil strife
Y_7 = social tension

$$Y_1 = \gamma_{10} + \beta_{12}Y_2 + \gamma_{12}X_2 + \gamma_{13}X_3 + u_1$$
$$Y_2 = \gamma_{20} + \beta_{21}Y_1 + \beta_{23}Y_3 + \gamma_{21}X_1 + u_2$$
$$Y_3 = \gamma_{30} + \beta_{32}Y_2 + \gamma_{32}X_2 + \gamma_{34}X_4 + \gamma_{35}X_5 + u_3$$
$$Y_4 = \gamma_{40} + \beta_{43}Y_3 + \beta_{45}Y_5 + \gamma_{42}X_2 + \gamma_{45}X_5 + u_4$$
$$Y_5 = \gamma_{50} + \beta_{52}Y_2 + \beta_{56}Y_6 + \gamma_{51}X_1 + u_5$$
$$Y_6 = \gamma_{60} + \beta_{61}Y_1 + \beta_{65}Y_5 + \beta_{67}Y_7 + \gamma_{61}X_1 + \gamma_{62}X_2 + u_6$$
$$Y_7 = (X_1 + Y_4)Y_3 \tag{11.1}$$

Note that we still keep to our convention of having only one variable on the left-hand side of the equation. Also notice that all the exogenous variables have an X symbol, and all the endogenous variables have a Y symbol. The coefficients β and γ are called structural coefficients. While β is always associated with endogenous variables, γ is always associated with exogenous variables. Finally, observe that the last equation in the system is not a regression equation but a determined relationship equation. The equation states unequivocally that Y_7 is obtained by multiplying the addition of X_1 and Y_4 by Y_3. There is no error term in this equation although all of the other stochastic equations have their own error term.

Because any particular variable can appear in any equation, we have to be able to identify not only its particular structural coefficient but also in which equation it appears. Thus each of the structural coefficients has two subscripts. The first indicates the number of the explained variable; the second identifies to which variable it is attached. Thus β_{21} tells us that this is the structural coefficient for endogenous variable Y_1 given that the explained variable is endogenous variable Y_2. The structural coefficients for the exogenous variables are similarly denoted. Thus γ_{34} locates our structural parameter for exogenous variable X_4, given that the explained variable in the equation is Y_3. This is the convention usually employed to explain the various difficulties with estimating these models. And it will be my policy to use this convention wherever possible. Our aim in these systems is to estimate the structural parameters β and γ.

11.2   The general form of a structural system

In order to analyze structural models more fully, we can put down the general form of a structural model.

$$\beta_{11}Y_{1t} + \beta_{12}Y_{2t} + \cdots + \beta_{1G}Y_{Gt} + \gamma_{10}X_{0t} + \gamma_{11}X_{1t} + \cdots + \gamma_{1K}X_{Kt} = u_{1t}$$

$$\beta_{21}Y_{1t} + \beta_{22}Y_{2t} + \cdots + \beta_{2G}Y_{Gt} + \gamma_{20}X_{0t} + \gamma_{21}X_{1t} + \cdots + \gamma_{1K}X_{Kt} = u_{2t}$$

$$\cdot \qquad \cdot \qquad \cdots \qquad \cdot \qquad \cdot \qquad \cdots \qquad \cdot \qquad \cdot$$

$$\beta_{G1}Y_{1t} + \beta_{G2}Y_{2t} + \cdots + \beta_{GG}Y_{Gt} + \gamma_{G0}X_{0t} + \gamma_{G1}X_{1t} + \cdots + \gamma_{GK}X_{Kt} = u_{Gt}$$

$$(11.2)$$

In this set of equations $X_{0t} = 1$. This is to accommodate the intercept term in each equation. In these equations all the Y's are endogenous variables and all the X's are exogenous variables. Thus we have G endogenous variables and $K + 1$ exogenous variables. Since we require an equation for every endogenous variable, we have G equations. The student should not confuse this K with the use of this symbol earlier in the text. (Previously it has been used as the total number of explained and explanatory variables in an equation. Obviously, K takes on a different meaning in the multiequation system.) The above set of equations is variously called a structural set of equations or a simultaneous set of equations. The subscript t refers to the observation number. For each variable we can have T observations where $t = 1, 2, 3, \ldots, T$.

In matrix formulation we can represent this whole set of equations as

$$\mathbf{B}\mathbf{Y}_t + \mathbf{\Gamma}\mathbf{X}_t = \mathbf{u}_t \qquad (11.3)$$

where

$$\mathbf{Y}_t = \begin{bmatrix} Y_{1t} \\ Y_{2t} \\ \cdot \\ Y_{Gt} \end{bmatrix} \qquad \mathbf{X}_t = \begin{bmatrix} X_{0t} \\ X_{1t} \\ \cdot \\ X_{Kt} \end{bmatrix} \qquad \mathbf{u}_t = \begin{bmatrix} u_{1t} \\ u_{2t} \\ \cdot \\ u_{Gt} \end{bmatrix}$$

$$(G \times 1) \qquad (K + 1 \times 1) \qquad (G \times 1)$$

$$\mathbf{B} = \begin{bmatrix} \beta_{11} & \beta_{12} & \cdots & \beta_{1G} \\ \beta_{21} & \beta_{22} & \cdots & \beta_{2G} \\ \cdot & \cdot & \cdots & \cdot \\ \beta_{G1} & \beta_{G2} & \cdots & \beta_{GG} \end{bmatrix} \qquad \mathbf{\Gamma} = \begin{bmatrix} \gamma_{10} & \gamma_{11} & \cdots & \gamma_{1K} \\ \gamma_{20} & \gamma_{21} & \cdots & \gamma_{2K} \\ \cdot & \cdot & \cdots & \cdot \\ \gamma_{G0} & \gamma_{G1} & \cdots & \gamma_{GK} \end{bmatrix}$$

$$(G \times G) \qquad\qquad (G \times K + 1)$$

At this point let me emphasize the assumptions about the error terms. First, we have the same assumptions concerned with each particular equation as we have had throughout the whole argument so far. That is, there should be no heteroscedasticity or autocorrelation of the error terms. The mean of the error term should also equal zero. More formally,

$$\text{cov}\,(u_{gt}, u_{gs}) = 0 \quad \text{where } u_{gt} \text{ is the error for observation } t \text{ in}$$
$$\text{equation } g$$
$$\text{and } u_{gs} \text{ is the error term for observation } s \text{ in}$$
$$\text{equation } g$$

and

$$u_{gt} \overset{\text{d}}{=} n(0, \sigma_{gg}^2)$$

However, across equations we do not make the assumption that error terms are independent as we did with recursive models. The error term for a particular case in one equation can be correlated with the error term for the same case in a different equation. More formally

$$\text{cov}\,(u_{gt},\,u_{ht}) \neq 0 \quad \text{where } u_{gt} \text{ is the error for observation } t \text{ in}$$

equation g

and u_{ht} is the error for observation t in equation h

Let us rearrange the conflict model of Gurr into this more general format, as shown on the facing page. From this format we can obtain the **B** and **Γ** matrices.

$$\mathbf{B} = \begin{bmatrix} 1 & -\beta_{12} & 0 & 0 & 0 & 0 & 0 \\ -\beta_{21} & 1 & -\beta_{23} & 0 & 0 & 0 & 0 \\ 0 & -\beta_{32} & 1 & 0 & 0 & 0 & 0 \\ 0 & 0 & -\beta_{43} & 1 & -\beta_{45} & 0 & 0 \\ 0 & -\beta_{52} & 0 & 0 & 1 & -\beta_{56} & 0 \\ -\beta_{61} & 0 & 0 & 0 & -\beta_{65} & 1 & -\beta_{67} \\ 0 & 0 & 0 & -Y_{3t} & 0 & 0 & 1 \end{bmatrix}$$

$$\mathbf{\Gamma} = \begin{bmatrix} -\gamma_{10} & 0 & -\gamma_{12} & -\gamma_{13} & 0 & 0 \\ -\gamma_{20} & -\gamma_{21} & 0 & 0 & 0 & 0 \\ -\gamma_{30} & 0 & -\gamma_{32} & 0 & -\gamma_{34} & -\gamma_{35} \\ -\gamma_{40} & 0 & -\gamma_{42} & 0 & 0 & -\gamma_{45} \\ -\gamma_{50} & -\gamma_{51} & 0 & 0 & 0 & 0 \\ -\gamma_{60} & -\gamma_{61} & -\gamma_{62} & 0 & 0 & 0 \\ 0 & -Y_{3t} & 0 & 0 & 0 & 0 \end{bmatrix}$$

We can see from the **B** matrix that there is a series of ones along the top-left to bottom-right diagonal. This is normal in most modeling situations because we generally select one of the endogenous variables to be the explained variable, The coefficient of this variable is set equal to one.

The ability to produce each structural set in this general form and hence obtain the matrices above is crucial in the processes that we will develop later. The student should not worry about the minus quantities in the matrices. They are due to the original formulation of the equations with

$$
\begin{aligned}
Y_{1t} - \beta_{12}Y_{2t} + 0\cdot Y_{3t} + 0\cdot Y_{4t} + 0\cdot Y_{5t} + 0\cdot Y_{6t} + 0\cdot Y_{7t} - \gamma_{10} + 0\cdot X_{1t} - \gamma_{12}X_{2t} - \gamma_{13}X_{3t} + 0\cdot X_{4t} + 0\cdot X_{5t} &= u_{1t} \\
-\beta_{21}Y_{1t} + Y_{2t} - \beta_{23}Y_{3t} + 0\cdot Y_{4t} + 0\cdot Y_{5t} + 0\cdot Y_{6t} + 0\cdot Y_{7t} - \gamma_{20} - \gamma_{21}X_{1t} + 0\cdot X_{2t} + 0\cdot X_{3t} + 0\cdot X_{4t} + 0\cdot X_{5t} &= u_{2t} \\
0\cdot Y_{1t} - \beta_{32}Y_{2t} + Y_{3t} + 0\cdot Y_{4t} + 0\cdot Y_{5t} + 0\cdot Y_{6t} + 0\cdot Y_{7t} - \gamma_{30} + 0\cdot X_{1t} - \gamma_{32}X_{2t} + 0\cdot X_{3t} - \gamma_{34}X_{4t} - \gamma_{35}X_{5t} &= u_{3t} \\
0\cdot Y_{1t} + 0\cdot Y_{2t} - \beta_{43}Y_{3t} + Y_{4t} - \beta_{45}Y_{5t} + 0\cdot Y_{6t} + 0\cdot Y_{7t} - \gamma_{40} + 0\cdot X_{1t} - \gamma_{42}X_{2t} + 0\cdot X_{3t} + 0\cdot X_{4t} - \gamma_{45}X_{5t} &= u_{4t} \\
0\cdot Y_{1t} - \beta_{52}Y_{2t} + 0\cdot Y_{3t} + 0\cdot Y_{4t} + Y_{5t} - \beta_{56}Y_{6t} + 0\cdot Y_{7t} - \gamma_{50} - \gamma_{51}X_{1t} + 0\cdot X_{2t} + 0\cdot X_{3t} + 0\cdot X_{4t} + 0\cdot X_{5t} &= u_{5t} \\
-\beta_{61}Y_{1t} + 0\cdot Y_{2t} + 0\cdot Y_{3t} + 0\cdot Y_{4t} - \beta_{65}Y_{5t} + Y_{6t} - \beta_{67}Y_{7t} - \gamma_{60} - \gamma_{61}X_{1t} - \gamma_{62}X_{2t} + 0\cdot X_{3t} + 0\cdot X_{4t} + 0\cdot X_{5t} &= u_{6t} \\
0\cdot Y_{1t} + 0\cdot Y_{2t} - Y_{3t}Y_{4t} + 0\cdot Y_{5t} + 0\cdot Y_{6t} + Y_{7t} + 0\cdot Y_{7t} + \gamma_{70} - Y_{3t}X_{1t} + 0\cdot X_{2t} + 0\cdot X_{3t} + 0\cdot X_{4t} + 0\cdot X_{5t} &= 0
\end{aligned}
\tag{11.4}
$$

the explained variable on the left and the explanatory variable on the right.

Any structural set of equations can be rearranged into the following matrix format:

$$[\mathbf{B}] \quad [\mathbf{Y}_t] \quad + \quad [\boldsymbol{\Gamma}] \quad [\mathbf{X}_t] \quad = \quad [\mathbf{u}_t]$$
$$(G \times G)(G \times 1) \quad (G \times K + 1)(K + 1 \times 1) \quad (G \times 1)$$

Another very important rearrangement of the structural set is into the reduced form.

11.3 The reduced form of a structural set of equations

It is possible to obtain each of the endogenous variables in terms of exogenous variables only. We can produce a set of equations, one for each of the endogenous variables, with the endogenous variables on the left-hand side of the equation and exogenous variables on the right-hand side. That is, we solve the structural-form equations for the endogenous variables. This produces the following system of equations:

$$Y_{1t} = \pi_{10}X_{0t} + \pi_{11}X_{1t} + \cdots + \pi_{1K}X_{Kt} + v_{1t}$$
$$Y_{2t} = \pi_{20}X_{0t} + \pi_{21}X_{1t} + \cdots + \pi_{2K}X_{Kt} + v_{2t}$$
$$\phantom{Y_{2t}} \cdot \qquad \cdot \qquad \cdot \qquad \qquad \cdot \qquad \qquad \cdot \tag{11.5}$$
$$Y_{Gt} = \pi_{G0}X_{0t} + \pi_{G1}X_{1t} + \cdots + \pi_{GK}X_{Kt} + v_{Gt}$$

In these equations the v's represent the reduced-form error terms and are some linear rearrangement of the structural error terms.

This set of equations can be encapsulated in the following matrix formulation:

$$\mathbf{Y}_t = \boldsymbol{\Pi}\mathbf{X}_t + \mathbf{v}_t \tag{11.6}$$

where:

$$\boldsymbol{\Pi} = \begin{bmatrix} \pi_{10} & \pi_{11} & \cdots & \pi_{1K} \\ \pi_{20} & \pi_{21} & \cdots & \pi_{2K} \\ \cdot & \cdot & & \cdot \\ \pi_{G0} & \pi_{G1} & \cdots & \pi_{GK} \end{bmatrix} \quad \mathbf{Y}_t = \begin{bmatrix} Y_{1t} \\ Y_{2t} \\ \cdot \\ Y_{Gt} \end{bmatrix} \quad \mathbf{X}_t = \begin{bmatrix} X_{0t} \\ X_{1t} \\ \cdot \\ X_{Kt} \end{bmatrix} \quad \mathbf{v}_t = \begin{bmatrix} v_{1t} \\ v_{2t} \\ \cdot \\ v_{Gt} \end{bmatrix}$$

In a pictorial fashion we see that

$$[\mathbf{Y}_t] \quad = \quad [\mathbf{\Pi}] \quad\quad [\mathbf{X}_t] \quad + \quad [\mathbf{v}_t]$$
$$(G \times 1) \quad (G \times K + 1)(K + 1 \times 1) \quad (G \times 1)$$

To help illustrate the processes, let us take a specific example. In Richardson's Arms Race theory the following two equations are provided:

A's arms level $= f$ (B's arms level, A's grievances about B)

B's arms level $= f$ (A's arms level, B's grievances about A)

In this equation set, A is one nation and B is another. We can put these into regression equation form producing the simultaneous set below.

$$Y_{1t} = \beta_{12}Y_{2t} + \gamma_{11}X_{1t} + u_{1t} \tag{11.7a}$$
$$Y_{2t} = \beta_{21}Y_{1t} + \gamma_{22}X_{2t} + u_{2t} \tag{11.7b}$$

where

Y_{1t} = arms level of nation A X_{1t} = A's grievances about B

Y_{2t} = arms level of nation B X_{2t} = B's grievances about A

This situation is pictured in diagram 11.2. From this we see that the grievance terms are exogenous variables, and that the model allows for no intercept term. Rearranging this into the general form of the structural set, we get

$$Y_{1t} - \beta_{12}Y_{2t} - \gamma_{11}X_{1t} + 0 \cdot X_{2t} = u_{1t}$$
$$-\beta_{21}Y_{1t} + \quad Y_{2t} + 0 \cdot X_{1t} - \gamma_{22}X_{2t} = u_{2t} \tag{11.8}$$

Diagram 11.2

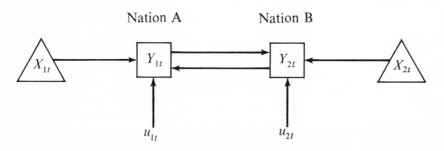

Nation A   Nation B

And further rearrangement into the matrix form produces

$$\begin{bmatrix} 1 & -\beta_{12} \\ -\beta_{21} & 1 \end{bmatrix}\begin{bmatrix} Y_{1t} \\ Y_{2t} \end{bmatrix} + \begin{bmatrix} -\gamma_{11} & 0 \\ 0 & -\gamma_{22} \end{bmatrix}\begin{bmatrix} X_{1t} \\ X_{2t} \end{bmatrix} = \begin{bmatrix} u_{1t} \\ u_{2t} \end{bmatrix}$$

To produce the reduced form we need to solve for the endogenous variables Y_{1t} and Y_{2t}. To do this we take the right-hand side of equation (11.7b) and insert it in place of Y_{2t} in equation (11.7a). In this way we obtain an equation for Y_{1t} in terms of the exogenous variables only.

$$Y_{1t} = \beta_{12}(\beta_{21}Y_{1t} + \gamma_{22}X_{2t} + u_{2t}) + \gamma_{11}X_{1t} + u_{1t}$$

$$Y_{1t} = \frac{\gamma_{11}}{1 - \beta_{12}\beta_{21}}X_{1t} + \frac{\beta_{12}\gamma_{22}}{1 - \beta_{12}\beta_{21}}X_{2t} + \frac{u_{1t} + \beta_{12}u_{2t}}{1 - \beta_{12}\beta_{21}} \qquad (11.9)$$

And putting this final equation back into equation (11.7b) we obtain Y_{2t} in terms of the exogenous variables only.

$$Y_{2t} = \beta_{21}\left(\frac{\gamma_{11}}{1 - \beta_{12}\beta_{21}} \cdot X_{1t} + \frac{\beta_{12}\gamma_{22}}{1 - \beta_{12}\beta_{21}} \cdot X_{2t} + \frac{u_{1t} + \beta_{12}u_{2t}}{1 - \beta_{12}\beta_{21}} \right)$$
$$+ \gamma_{22}X_{2t} + u_{2t}$$

$$Y_{2t} = \frac{\beta_{21}\gamma_{11}}{1 - \beta_{12}\beta_{21}} \cdot X_{1t} + \frac{\gamma_{22}}{1 - \beta_{12}\beta_{21}} \cdot X_{2t} + \frac{\beta_{21}u_{1t} + u_{2t}}{1 - \beta_{12}\beta_{21}} \qquad (11.10)$$

Thus we have the reduced form of the system. Notice how the endogenous variables are on the left-hand side of the equation while the exogenous variables are on the right-hand side. When we put equations (11.9) and (11.10) into the matrix form of the reduced form we obtain

$$\begin{bmatrix} Y_{1t} \\ Y_{2t} \end{bmatrix} = \begin{bmatrix} \dfrac{\gamma_{11}}{1 - \beta_{12}\beta_{21}} & \dfrac{\beta_{12}\gamma_{22}}{1 - \beta_{12}\beta_{21}} \\ \dfrac{\beta_{21}\gamma_{11}}{1 - \beta_{12}\beta_{21}} & \dfrac{\gamma_{22}}{1 - \beta_{12}\beta_{21}} \end{bmatrix}\begin{bmatrix} X_{1t} \\ X_{2t} \end{bmatrix} + \begin{bmatrix} \dfrac{u_{1t} + \beta_{12}u_{2t}}{1 - \beta_{12}\beta_{21}} \\ \dfrac{\beta_{21}u_{1t} + u_{2t}}{1 - \beta_{12}\beta_{21}} \end{bmatrix}$$

Thus we have now produced the arms-race equations in two forms, the structural form given by the matrix equation,

$$\mathbf{B}\mathbf{Y}_t + \mathbf{\Gamma}\mathbf{X}_t = \mathbf{u}_t$$

and the reduced form given by

$$\mathbf{Y}_t = \mathbf{\Pi}\mathbf{X}_t + \mathbf{v}_t$$

We can very quickly show a relationship between the two forms. We can rearrange the structural equation as below.

$$\mathbf{BY}_t = -\mathbf{\Gamma X}_t + \mathbf{u}_t \tag{11.11}$$

If we premultiply by $\mathbf{B}^{-1}$ we obtain

$$\mathbf{Y}_t = -\mathbf{B}^{-1}\mathbf{\Gamma X}_t + \mathbf{B}^{-1}\mathbf{u}_t$$

and when we compare this relationship with the reduced form we see that

$$\mathbf{\Pi} = -\mathbf{B}^{-1}\mathbf{\Gamma} \qquad \text{and} \qquad \mathbf{v}_t = \mathbf{B}^{-1}\mathbf{u}_t \tag{11.12}$$

Let us check the first of these relationships using the arms-race example. The inversion of the $\mathbf{B}$ matrix follows the pattern

$$\mathbf{B} = \begin{bmatrix} 1 & -\beta_{12} \\ -\beta_{21} & 1 \end{bmatrix}$$

Determinant $|\mathbf{B}| = [1 - \beta_{12}\beta_{21}]$

Cofactor for $\mathbf{B} = \begin{bmatrix} 1 & \beta_{21} \\ \beta_{12} & 1 \end{bmatrix}$

Adjoint $\mathbf{B} = \begin{bmatrix} 1 & \beta_{12} \\ \beta_{21} & 1 \end{bmatrix}$

$$\mathbf{B}^{-1} = \begin{bmatrix} \dfrac{1}{1 - \beta_{12}\beta_{21}} & \dfrac{\beta_{12}}{1 - \beta_{21}\beta_{12}} \\[2ex] \dfrac{\beta_{21}}{1 - \beta_{12}\beta_{21}} & \dfrac{1}{1 - \beta_{12}\beta_{21}} \end{bmatrix}$$

and multiplying that by $-\mathbf{\Gamma}$,

$$\mathbf{\Pi} = -\mathbf{B}^{-1}\mathbf{\Gamma} = -\begin{bmatrix} \dfrac{1}{1 - \beta_{12}\beta_{21}} & \dfrac{\beta_{12}}{1 - \beta_{21}\beta_{12}} \\[2ex] \dfrac{\beta_{21}}{1 - \beta_{12}\beta_{21}} & \dfrac{1}{1 - \beta_{12}\beta_{21}} \end{bmatrix} \begin{bmatrix} -\gamma_{11} & 0 \\[2ex] 0 & -\gamma_{22} \end{bmatrix}$$

$$= \begin{bmatrix} \dfrac{\gamma_{11}}{1 - \beta_{12}\beta_{21}} & \dfrac{\beta_{12}\gamma_{22}}{1 - \beta_{21}\beta_{12}} \\[2ex] \dfrac{\gamma_{11}\beta_{21}}{1 - \beta_{12}\beta_{21}} & \dfrac{\gamma_{22}}{1 - \beta_{12}\beta_{21}} \end{bmatrix}$$

This relationship between the structural equation and the reduced-form equation is crucial in the analysis that follows.

11.4   Some problems with the structural systems

As we have seen, we cannot treat the different equations of any structural set as individual equations. This is because the endogenous variables are common throughout the structural set. Any particular endogenous variable may appear in several equations in the structural set simultaneously, thereby linking the equations. Estimation processes for the structural coefficient of such a variable in one equation cannot be carried out without reference to the presence of that variable in another equation.

Put bluntly, the estimates of the coefficients are inconsistent. This we will show shortly. But more than this, there may be no consistent estimate of the coefficient. That is, the model is not identified correctly.

Let us use our arms-race example. If we were to plot the arms of nation A against the arms of nation B on a scattergram and draw a regression line through the points we would get diagram 11.3. As the diagram shows, there is an equilibrium point, E. At this point, the arms level of nation A, Y_{1t} is equal in both the arms-level equation for A and the arms-level equation for nation B. From equations (11.7a) and (11.7b), we can put down

Arms-level line for nation A $Y_{1t} = \beta_{12} Y_{2t} + \gamma_{11} X_{1t} + u_{1t}$

Arms-level line for nation B $Y_{1t} = \dfrac{1}{\beta_{21}} \cdot Y_{2t} - \dfrac{\gamma_{22}}{\beta_{21}} \cdot X_{2t} - \dfrac{1}{\beta_{21}} \cdot u_{2t}$

From these we can produce

$$\beta_{12} Y_{2t} + \gamma_{11} X_{1t} + u_{1t} = \dfrac{1}{\beta_{21}} Y_{2t} - \dfrac{\gamma_{22}}{\beta_{21}} X_{2t} - \dfrac{1}{\beta_{21}} u_{2t}$$

and rearranging this, we obtain

$$Y_{2t} = \dfrac{\beta_{21}\gamma_{11}}{1 - \beta_{12}\beta_{21}} X_{1t} + \dfrac{\gamma_{22}}{1 - \beta_{12}\beta_{21}} X_{2t} + \dfrac{\beta_{21}u_{1t} + u_{2t}}{1 - \beta_{12}\beta_{21}}$$

Notice that this is a reduced-form equation we established earlier. But in addition, notice that Y_{2t} has the term u_{1t} in the error term on the right-hand side of the equation. Thus Y_{2t} is not independent of u_{1t}, and if we were to

Diagram 11.3

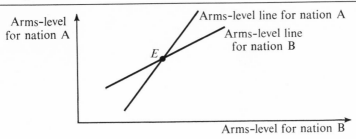

estimate the coefficients in equation (11.7a) using ordinary least-squares where Y_{2t} is an explanatory variable and u_{1t} is the error term, we would have contemporaneous correlation between an explanatory variable and the error term in the same equation. As we saw in Chapter 5 this will produce inconsistent estimates of the structural coefficients. And no increase in the sample size will remedy this situation.

11.5   The identification problem

In dealing with the simultaneous equations that make up a structural system, we encounter one of the most difficult and frustrating problems confronting the politometrician. The problem is the identification problem and ironically is not a statistical but a mathematical problem. Although we shall look at identification in different ways, the student can obtain some preliminary idea of the difficulty by reconsidering elementary algebra knowledge. Given any set of simultaneous equations, no solution is possible unless there are at least as many equations as there are unknowns in the equations. If there are more unknowns than equations, then no solution exists. If there are more equations than unknowns, then too many solutions exist. Only if there are exactly the same number of unknowns as equations can we obtain a unique solution for the unknowns. This is the identification problem in its simplest form. The student will quickly see the relevance since in our structural model we have both equations and unknowns.

In this remaining part of this chapter, I intend to consider the identification problem from various viewpoints using diagrams. I can assure students that the mathematics are quite within their capabilities, assuming they have gotten this far in the text. No new mathematical ideas are introduced, and a careful study of this chapter and the next will equip any student with the knowledge to conquer, where it is possible, the identification problem.

For the student not prepared to go through the discussion of the problem, I will provide rules to determine whether equations are identified and will present some additional ideas on how to get out of situations where identification is a problem.

11.5.1   A first look at the problem

For the sake of this argument let us change the arms-race equations to exclude the grievance terms. Pictorially this would be diagram 11.4. The formal model is

$$Y_{1t} = \beta_{12} Y_{2t} + \gamma_{10} X_{0t} \quad \text{Nation A's arms-level line}$$
$$Y_{2t} = \beta_{21} Y_{1t} + \gamma_{20} X_{0t} \quad \text{Nation B's arms-level line}$$

Diagram 11.4

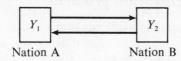

Nation A   Nation B

Notice that there are no error terms in this system. At the equilibrium point for the system, if we obtained observations for the arms level of both nations each year, all our observations would be exactly the same. If we were to plot these observations on a scattergram, all we would obtain is one data point (diagram 11.5). The student will quickly see that we cannot locate the two regression lines on this scattergram because

Diagram 11.5

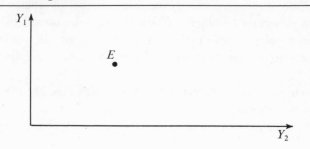

there just is not enough information there. There is an infinity of different ways we can draw two lines through this one point. The model is not identified.

Let us now add an error term and therefore some variance into the model. The model would look like that shown in diagram 11.6. Mathematically this would be

$$Y_{1t} = \beta_{12}Y_{2t} + \gamma_{10}X_{0t} + u_{1t} \quad \text{Nation A's line}$$
$$Y_{2t} = \beta_{21}Y_{1t} + \gamma_{20}X_{0t} + u_{2t} \quad \text{Nation B's line}$$

Diagram 11.6

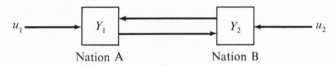

Nation A   Nation B

This error term will allow the lines, which we do not know, to move horizontally parallel in the case of nation B's line and vertically parallel in the case of nation A's line. Diagram 11.7 demonstrates this. We have selected hypothetical slopes for the two lines and just moved these parallel as indicated by potential error-term changes. But if we take away the lines, all we have left is what the politometrician gets, a scatterplot. The student will see that it is still impossible to obtain the true regression lines from these points. Estimates obtained from these points would be entirely erroneous. We see that we still cannot identify the model.

Suppose now that we reintroduce the grievance term into one of the equations, let's say as influencing the behavior of nation A. At the same

Diagram 11.7

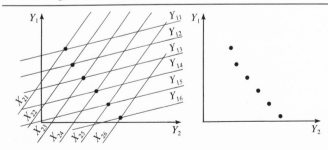

time let us eliminate the error term in both equations. The model would become diagram 11.8, which in mathematical terms would be

$$Y_{1t} = \beta_{12}Y_{2t} + \gamma_{10}X_{0t} + \gamma_{11}X_{1t} \quad \text{Arms level for nation A}$$
$$Y_{2t} = \beta_{21}Y_{1t} + \gamma_{20}X_{0t} \quad \text{Arms level for nation B}$$

Diagram 11.8

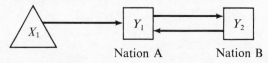

Nation A   Nation B

Suppose further that we let the grievance term in the nation A equation vary quite considerably over time. Because of this, the slope of that line for the arms level of nation A would remain constant but would move up and down the scatterplot in a parallel motion. Notice, however, that the line for the arms level of nation B will remain constant. Thus where these lines crossed would produce a series of observation plots. On the left in diagram 11.9 we have this omniscient ability to locate the lines and establish the various points of intersection, but on the right we have the real situation confronting the politometrician, a series of data plots.

Diagram 11.9

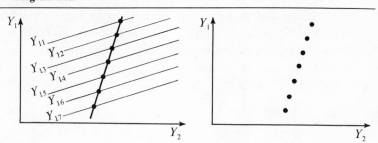

Notice that the data points in the sketch on the right actually map out the line for the arms level of nation B. Thus we can identify the line for the arms level of nation B. And further notice that it is an increase in the exogenous variables in the line for the arms level of nation A that has made this possible. We shall later remember this when looking for clues to determine whether a particular line is identified or not. All we notice at this stage is that we need an extra exogenous variable in one of the equations, and further that this exogenous variable must be allowed to vary so that the line might move in a parallel fashion.

Let us make the alternative assumption concerning the grievance term of nation B. The model would be as in diagram 11.10. And mathematically,

$$Y_{1t} = \beta_{12}Y_{2t} + \gamma_{10}X_{0t} \qquad \text{Arms level for nation A}$$

$$Y_{2t} = \beta_{21}Y_{1t} + \gamma_{20}X_{0t} + \gamma_{22}X_{2t} \quad \text{Arms level for nation B}$$

Diagram 11.10

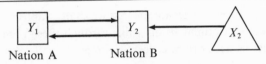

Nation A   Nation B

And as with the previous situation, if we let the grievance term vary we will cause the line for nation B to move in a horizontal fashion while the line for nation A will remain still. Diagram 11.11 shows how the equilibrium points are formed. The sketch on the right more particularly shows that this time we have identified the line for nation A by allowing the exogenous variable in the line for nation B to vary.

Diagram 11.11

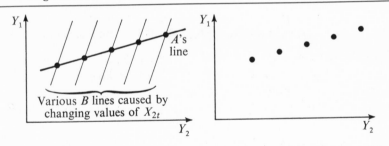

This concludes our first look at the problem of identification. The student should notice that identification is not possible at all in models without exogenous variables. The student should further notice that the placing of a varying exogenous variable in one equation allows other equations to become identified. If we want to, we can add the error terms to either of the equations. All this would do is to slightly diffuse the mapping out of the line. Instead of getting a perfectly straight line for the identified line, we would obtain something like that presented in diagram

Diagram 11.12

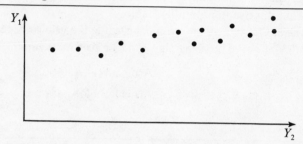

11.12. The least-squares regression technique that we shall establish in the next chapter will locate an optimal straight line through these points in much the same manner as does ordinary least-squares.

11.5.2 A second look at the problem

Let us now approach the same problem but from a different viewpoint. The argument presented here should not be seen as following from the previous argument but as complementary to it. Indeed, the more viewpoints we get, the likelier it is that we shall understand the problem.

To help us here we will use the arms-race models again without the exogenous variables but including the error terms.

$$Y_{1t} = \beta_{12}Y_{2t} + \gamma_{10}X_{0t} + u_{1t} \quad \text{Arms-level line, nation A}$$
$$Y_{2t} = \beta_{21}Y_{1t} + \gamma_{20}X_{0t} + u_{2t} \quad \text{Arms-level line, nation B}$$

Further we will assume that u_{1t} and u_{2t} are independent of one another. The randomness of the error term will, if the sample size is very large, produce a scattergram similar to that shown in diagram 11.13. The distribution will be elliptical, and the shape and disposition of the ellipse will

Diagram 11.13

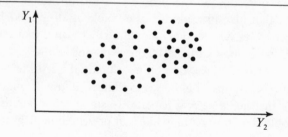

be a function of the slopes of the two regression lines and the variance of the error terms.

The problem confronting the politometrician is, Given these data points, how can he locate the two true regression lines underlying the generation of the data points? The only information is that the error terms are unrelated and that they both have a mean of zero. Since this is the case, the two lines must divide the area of the ellipse into four equal parts. If it were otherwise, we could not conclude that the error terms were unrelated. Consider diagram 11.14. Obviously, in this situation, when both Y_{1t} and Y_{2t} have high values the error terms are also high. Conversely, when these variables have low values, their errors are low. Thus the error terms are related.

Diagram 11.14

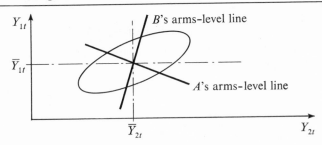

However, we know that our error terms are not related because this is one of our assumptions. Thus we know that the two regression lines must divide the ellipse area into four equal parts. Unfortunately this is all we do know. And there are an infinity of pairs of lines which will satisfy this restriction. The problem becomes one of selection among all possible pairs of lines which satisfy this restriction. We do not have enough information to identify the lines.

So it is a matter of added information. The next question becomes, What kind of information? Let us look at two kinds of prior information that would help.

Suppose we know that there is no error in our measurement of Y_{1t}. That is, the u_{1t} is equal to zero, or more plausibly, the error in measuring Y_{1t} is very much less than that in measuring Y_{2t}. We would get the plot shown in diagram 11.15. The thinness of the ellipse drastically reduces the number of pairs of lines which divide the ellipse into four equal parts. In

Diagram 11.15

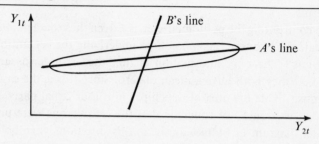

particular, the number of possible lines for the arms level of nation A is very small, allowing us to obtain a reasonable estimate of that line. Our assumptions in this example are that the two error terms are unrelated and that one of the error distributions is very small compared with the other.

Another kind of prior information might concern the value of one of the regression slopes. Suppose, for example, that for line Y_{2t} we have determined the value of β_{21}. From our previous discussions of regression we know that this line will go through the middle of the ellipse, that is, through the intersection of the mean of Y_{1t} and Y_{2t}. We can thus locate the line quite accurately with this information. But we also know that the error terms are unrelated and that the pair of lines must divide the ellipse area exactly into four equal parts. With a very large sample size there is only one Y_{1t} line that will actually interact with the established Y_{2t} line to quarter the ellipse area (diagram 11.16). In this situation two portions of prior knowledge—namely, the independence of the error terms and the value of the slope of one of the lines—have led to an identification of the

Diagram 11.16

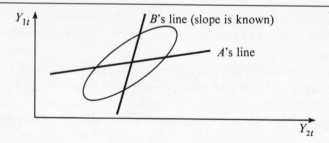

equations. We see that our situation is improved with additional information, but we will also see that too much information produces other problems, less severe than those already encountered, but nevertheless something to be concerned about.

Before we continue on to a more mathematical analysis of the identification problem, let us summarize what we have seen so far. First, we have seen that it is a matter of information shortage. In certain circumstances, there is just not enough information in the system of equations to enable us to arrive at a solution. It is a mathematical problem rather than a statistical problem. Indeed, if we were to increase the sample size under the mistaken conclusion that it was a statistical problem, then all we would do is make ourselves more confident about erroneous estimates. We have also seen that we can often get out of the difficulty by making various assumptions or searching for prior information.

Further readings

Substantive
Choucri, N., and R. C. North. 1975. *Nations in Conflict*. San Francisco: Freeman.
Gurr, T. R., and R. Duvall. 1973. "Civil Conflict in the 1960's: A Reciprocal Theoretical System with Parameter Estimates." *Comparative Political Studies* 6: 135.
Hibbs, D. A. 1973. *Mass Political Violence: A Cross-National Causal Analysis*. New York: Wiley.

Statistical
Johnston: pp. 352–75.
Kmenta: pp. 531–49.
Wonnacott and Wonnacott: chapters 8, 18.

12

the identification problem and parameter estimation

In this final chapter we shall consider a more rigorous analysis of the identification problem and eventually provide rules by which the politometrician can determine whether any particular equation in a structural set is identified. Having done all of this and assuming an equation is identified, we require estimation techniques. These are also provided in this chapter. And in a final section we shall deal with miscellaneous structural models which might be found in political science. But first the mathematical analysis of identification.

12.1　A mathematical analysis of the identification problem

In Chapter 11 we saw that regression analysis of equations within a structural system would produce inconsistent estimates of the regression coefficients. However, if we convert the equations into the reduced form, we would have endogenous variables on the left-hand side of the equations and exogenous variables on the right-hand side of the equations. We would, on regression of the reduced form, obtain consistent estimates of the coefficients π.

But we do not really want these coefficients; we want the structural coefficients β and γ, and we know that they are all related by the matrix equation

$$\Pi = -B^{-1}\Gamma$$

It would seem an easy matter to solve this equation for the matrices **B** and **Γ**, knowing **Π**. But it is not that easy! There has to be exactly the same amount of information in the **Π** matrix as that required by the **B** and **Γ** matrices. As we mentioned previously, it becomes a matter of the number of unknowns compared to the number of equations. We can define three conditions in the relationship of the number of unknowns to the number of equations available.

(1) If there are more unknowns than equations we have an *under-identified* model.
(2) If there are exactly the same number of unknowns as there are equations we have an *identified* model.
(3) If there are more equations than unknowns then we have an *overidentified* model.

Condition (1) leaves us impotent unless we can come up with some added information. Condition (2) is very agreeable, giving us exactly the amount of information for unique solutions. Condition (3) provides us with too many solutions. Condition (2) is to be preferred to condition (3), which is to be preferred to condition (1).

Let us fabricate another arms-race problem to help clarify the problem. This time we will allow a triadic relationship with nations A, B, and C. The model is

$$Y_{1t} = \beta_{12}Y_{2t} + \gamma_{10}X_{0t} + u_{1t} \qquad \text{nation A's line} \quad (12.1a)$$

$$Y_{2t} = \beta_{21}Y_{1t} + \gamma_{20}X_{0t} + \gamma_{22}X_{2t} + u_{2t} \quad \text{nation B's line} \quad (12.1b)$$

$$Y_{3t} = \beta_{31}Y_{1t} + \beta_{32}Y_{2t} + \gamma_{30}X_{0t} + \gamma_{33}X_{3t} + u_{3t}$$
$$\text{nation C's line} \quad (12.1c)$$

In these equations,

Y_{1t} = arms level for nation A
Y_{2t} = arms level for nation B
Y_{3t} = arms level for nation C
X_{2t} and X_{3t} = grievance terms for nations B and C respectively
X_{0t} = intercept term
u_{1t}, u_{2t}, and u_{3t} = error terms

Schematically the model is as shown in diagram 12.1. The reduced-form

Diagram 12.1

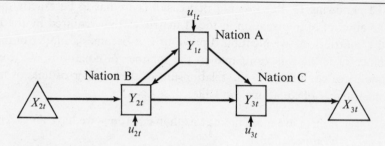

equations for this structural set are:

$$Y_{1t} = \pi_{10}X_{0t} + \pi_{12}X_{2t} + \pi_{13}X_{3t} + v_{1t} \qquad (12.2a)$$

$$Y_{2t} = \pi_{20}X_{0t} + \pi_{22}X_{2t} + \pi_{23}X_{3t} + v_{2t} \qquad (12.2b)$$

$$Y_{3t} = \pi_{30}X_{0t} + \pi_{32}X_{2t} + \pi_{33}X_{3t} + v_{3t} \qquad (12.2c)$$

We are now in a position to determine the identifiability status of each of the equations in turn. Each equation has to be considered individually. Let us look at the line for the arms level of nation A.

$$Y_{1t} = \beta_{12}Y_{2t} + \gamma_{10}X_{0t} + u_{1t} \qquad \text{structural form}$$

also

$$Y_{1t} = \pi_{10}X_{0t} + \pi_{12}X_{2t} + \pi_{13}X_{3t} + v_{1t} \quad \text{reduced form}$$

thus

$$\pi_{10}X_{0t} + \pi_{12}X_{2t} + \pi_{13}X_{3t} + v_{1t} = \beta_{12}Y_{2t} + \gamma_{10}X_{0t} + u_{1t}$$

but

$$Y_{2t} = \pi_{20}X_{0t} + \pi_{22}X_{2t} + \pi_{23}X_{3t} + v_{2t}$$

Inserting the equation in the previous one, we obtain

$$\pi_{10}X_{0t} + \pi_{12}X_{2t} + \pi_{13}X_{3t} + v_{1t}$$
$$= \beta_{12}(\pi_{20}X_{0t} + \pi_{22}X_{2t} + \pi_{23}X_{3t} + v_{2t}) + \gamma_{10}X_{0t} + u_{1t}$$

All of the endogenous variables are thus eliminated from the equation. Solving, we get

$$\pi_{10}X_{0t} + \pi_{12}X_{2t} + \pi_{13}X_{3t} + v_{1t}$$
$$= (\beta_{12}\pi_{20} + \gamma_{10})X_{0t} + \beta_{12}\pi_{22}X_{2t} + \beta_{12}\pi_{23}X_{3t} + \beta_{12}v_{2t} + u_{1t}$$

Comparing similar terms on either side of the equation, we obtain the equalities

$$\pi_{10} = \beta_{12}\pi_{20} + \gamma_{10}$$
$$\pi_{12} = \beta_{12}\pi_{22}$$
$$\pi_{13} = \beta_{12}\pi_{23}$$

We will have obtained values for all the π coefficients by carrying out OLS on the reduced form of the equation. Now the trick is to get estimates of the structural coefficients β_{12} and γ_{10}, given that we have π_{10}, π_{12}, π_{13}, π_{20}, π_{22}, and π_{23} form OLS on the reduced form. This is where we need to compare the number of unknowns with the number of equations. We have two unknowns, β_{12} and γ_{10}, and three equations. Thus we have an *overidentified* equation, and we will obtain more than one solution; for instance $\beta_{12} = \pi_{13}/\pi_{23}$ or π_{12}/π_{22}. We conclude that our equation for the arms level of nation A is overidentified.

Let us now look at the arms level equation for nation B. The equation is

$$Y_{2t} = \beta_{21}Y_{1t} + \gamma_{20}X_{0t} + \gamma_{22}X_{2t} + u_{2t}$$

The reduced-form equation is

$$Y_{2t} = \pi_{20}X_{0t} + \pi_{22}X_{2t} + \pi_{23}X_{3t} + v_{2t}$$

Equating these two, we obtain

$$\pi_{20}X_{0t} + \pi_{22}X_{2t} + \pi_{23}X_{3t} + v_{2t} = \beta_{21}Y_{1t} + \gamma_{20}X_{0t} + \gamma_{22}X_{2t} + u_{2t}$$

Replacing Y_{1t} in this equation by

$$\pi_{10}X_{0t} + \pi_{12}X_{2t} + \pi_{13}X_{3t} + v_{1t}$$

we get

$$
\begin{aligned}
\pi_{20}X_{0t} &+ \pi_{22}X_{2t} + \pi_{23}X_{3t} + v_{2t} \\
&= \beta_{21}(\pi_{10}X_{0t} + \pi_{12}X_{2t} + \pi_{13}X_{3t} + v_{1t}) \\
&\quad + \gamma_{20}X_{0t} + \gamma_{22}X_{2t} + u_{2t} \\
&= (\beta_{21}\pi_{10} + \gamma_{20})X_{0t} + (\beta_{21}\pi_{12} + \gamma_{22})X_{2t} \\
&\quad + \beta_{21}\pi_{13}X_{3t} + \beta_{21}v_{1t} + u_{2t}
\end{aligned}
$$

When we compare similar terms on either side of the equality, we obtain

$$\pi_{20} = \beta_{21}\pi_{10} + \gamma_{20}$$
$$\pi_{22} = \beta_{21}\pi_{12} + \gamma_{22}$$
$$\pi_{23} = \beta_{21}\pi_{13}$$

In this situation we have three unknowns, β_{21}, γ_{20}, and γ_{22}, and three equations. We thus have the *exactly identified* condition in this equation for the arms level of nation B.

Finally, let us look at the line for the arms level of nation C.

$$Y_{3t} = \beta_{31}Y_{1t} + \beta_{32}Y_{2t} + \gamma_{30}X_{0t} + \gamma_{33}X_{3t} + u_{3t}$$

The reduced form for this is

$$Y_{3t} = \pi_{30}X_{0t} + \pi_{32}X_{2t} + \pi_{33}X_{3t} + v_{3t}$$

Equating these two equations we obtain:

$$\pi_{30}X_{0t} + \pi_{32}X_{2t} + \pi_{33}X_{3t} + v_{3t}$$
$$= \beta_{31}Y_{1t} + \beta_{32}Y_{2t} + \gamma_{30}X_{0t} + \gamma_{33}X_{3t} + u_{3t}$$

and on substituting reduced-form equations for Y_{1t} and Y_{2t}, we get

$$\pi_{30}X_{0t} + \pi_{32}X_{2t} + \pi_{33}X_{3t} + v_{3t}$$
$$= \beta_{31}(\pi_{10}X_{0t} + \pi_{12}X_{2t} + \pi_{13}X_{3t} + v_{1t})$$
$$+ \beta_{32}(\pi_{20}X_{0t} + \pi_{22}X_{2t} + \pi_{23}X_{3t} + v_{2t})$$
$$+ \gamma_{30}X_{0t} + \gamma_{33}X_{3t} + u_{3t}$$
$$= (\beta_{31}\pi_{10} + \beta_{32}\pi_{20} + \gamma_{30})X_{0t} + (\beta_{31}\pi_{12} + \beta_{32}\pi_{22})X_{2t}$$
$$+ (\beta_{31}\pi_{13} + \beta_{32}\pi_{23} + \gamma_{33})X_{3t}$$
$$+ \beta_{31}v_{1t} + \beta_{32}v_{2t} + u_{3t}$$

When we equate similar terms on either side of the equation, we obtain the three equalities

$$\pi_{30} = \beta_{31}\pi_{10} + \beta_{32}\pi_{20} + \gamma_{30}$$
$$\pi_{32} = \beta_{31}\pi_{12} + \beta_{32}\pi_{22}$$
$$\pi_{33} = \beta_{31}\pi_{13} + \beta_{32}\pi_{23} + \gamma_{33}$$

In these equations we have four unknowns, β_{31}, β_{32}, γ_{30}, and γ_{33}, and only three equations. Thus we cannot obtain estimates of the four unknowns and conclude that this equation is *underidentified*.

We have therefore determined the identifiability status of the three equations in the arms-race model. Notice that we can always get estimates for the reduced-form coefficients. However, we require the structural coefficients and need to determine these using the reduced-form estimates. But we have seen that there can be varying amounts of information in the reduced-form coefficients—sometimes too much (overidentification), sometimes too little (underidentification), and sometimes just enough (exact identification). The student should notice that all the above arguments have been made without recourse to any data.

12.2   Rules for determining identification status

Obviously it would be desirable to develop some rules for determining at a glance whether a particular equation in a given structural set is underidentified, exactly identified, or overidentified. This would eliminate the need to go through the process described in the previous section. What follows is a mathematical generation of such rules. The determined student can work his or her way through the argument and will, I think, derive much pleasure from the effort. The more indolent, or should I say pragmatic, student can skip to the end of the section where the rules for determining the identification status of equations in structural systems can be found. My aim in this section is to develop the formal conditions for identification.

From equation (11.12) we have

$$\Pi = -\mathbf{B}^{-1}\Gamma$$

And so

$$\mathbf{B}\Pi = -\Gamma \tag{12.3}$$

Putting this back into matrix form, we get

$$
\begin{bmatrix}
\beta_{11} & \beta_{12} & \beta_{13} & \cdots & \beta_{1G} \\
\beta_{21} & \beta_{22} & \beta_{23} & \cdots & \cdot \\
\cdot & & & & \cdot \\
\beta_{G1} & \beta_{G2} & \beta_{G3} & \cdots & \beta_{GG}
\end{bmatrix}
\begin{bmatrix}
\pi_{10} & \pi_{11} & \pi_{12} & \cdots & \pi_{1K} \\
\pi_{20} & & & & \cdot \\
\cdot & & & & \cdot \\
\pi_{G0} & \cdot & \cdot & \cdots & \pi_{GK}
\end{bmatrix}
$$

$$(G \times G) \qquad\qquad (G \times K + 1)$$

$$
= -
\begin{bmatrix}
\gamma_{10} & \gamma_{11} & \gamma_{12} & \cdots & \gamma_{1K} \\
\gamma_{20} & & & & \cdot \\
\cdot & & & & \cdot \\
\gamma_{G0} & \cdot & \cdot & \cdots & \gamma_{GK}
\end{bmatrix}
$$

$$(G \times K + 1)$$

But we intend to scrutinize one equation at a time. Let us for the sake of generality select the gth equation. This will be given by

$$\beta_g \Pi = -\gamma_g \tag{12.4}$$

In matrix form this is

$$
[\beta_{g1} \quad \beta_{g2} \quad \beta_{g3} \quad \cdots \quad \beta_{gG}]
\begin{bmatrix}
\pi_{10} & \pi_{11} & \pi_{12} & \cdots & \pi_{1K} \\
\pi_{20} & & & & \cdot \\
\cdot & & & & \cdot \\
\pi_{G0} & \cdot & \cdot & \cdots & \pi_{GK}
\end{bmatrix}
$$

$$(1 \times G)$$
$$(G \times K + 1)$$

$$
= -[\gamma_{g0} \quad \gamma_{g1} \quad \gamma_{g2} \quad \cdots \quad \gamma_{gK}]
$$
$$(1 \times K + 1)$$

We can assume that some of the β's and γ's are equal to zero. That is, some variables are not included in the equation. We can therefore define the following:

G^Δ = number of endogenous variables appearing in the gth equation

$G^{\Delta\Delta} = G - G^\Delta$ = number of endogenous variables left out of the equation

K^* = number of exogenous variables that appear in the gth equation

$K^{**} = K + 1 - K^*$ = number of exogenous variables left out of the equation

Using the partitioning trick we learned in Chapter 6, we can rearrange the gth equation with the nonzero elements of β_g and γ_g first. Thus we can put down

$$\beta_g = [\boldsymbol{\beta}_\Delta \mid \mathbf{0}_{\Delta\Delta}]$$
$$\gamma_g = [\boldsymbol{\gamma}_* \mid \mathbf{0}_{**}] \tag{12.5}$$

where

$\boldsymbol{\beta}_\Delta = [\beta_{g1} \quad \cdots \quad \beta_{gG}\Delta]$ a $1 \times G^\Delta$ row vector

$\mathbf{0}_{\Delta\Delta} = [0 \quad \cdots \quad 0]$ a $1 \times G^{\Delta\Delta}$ row vector

$\boldsymbol{\gamma}_* = [\gamma_{g0} \quad \cdots \quad \gamma_{gK}^*]$ a $1 \times K^*$ row vector

$\mathbf{0}_{**} = [0 \quad \cdots \quad 0]$ a $1 \times K^{**}$ row vector

We can go even further and partition the Π matrix so that it remains conformable with β_g and γ_g. Partitioning produces

$$\Pi = \left[\begin{array}{c|c} \pi_{\Delta *} & \pi_{\Delta **} \\ \hline \pi_{\Delta\Delta *} & \pi_{\Delta\Delta **} \end{array}\right] \tag{12.6}$$

where

$\pi_{\Delta *}$ is a $G^\Delta \times K^*$ matrix

$\pi_{\Delta **}$ is a $G^\Delta \times K^{**}$ matrix

$\pi_{\Delta\Delta *}$ is a $G^{\Delta\Delta} \times K^*$ matrix

$\pi_{\Delta\Delta **}$ is a $G^{\Delta\Delta} \times K^{**}$ matrix

Putting all of these partitions back into equation (12.4), we obtain

$$[\beta_\Delta \mid 0_{\Delta\Delta}] \left[\begin{array}{c|c} \pi_{\Delta *} & \pi_{\Delta **} \\ \hline \pi_{\Delta\Delta *} & \pi_{\Delta\Delta **} \end{array}\right] = -[\gamma_* \quad 0_{**}] \tag{12.7}$$

This leads us to the following equations:

$$\beta_\Delta \pi_{\Delta *} = -\gamma_* \tag{12.8}$$
$$(1 \times G^\Delta)(G^\Delta \times K^*) \quad (1 \times K^*)$$

$$\beta_\Delta \pi_{\Delta **} = 0_{**} \tag{12.9}$$
$$(1 \times G^\Delta)(G^\Delta \times K^{**}) \quad (1 \times K^{**})$$

We can easily solve equation (12.9) for β_Δ, since γ is not involved in the equation. Notice that this will have K^{**} equations.

The next question is, How many unknowns are there in this equation set? The only unknowns are the β's, and we have G^Δ of them. But we know that one of the β's is unity since we always let one of the endogenous variables be the explained variable. So in reality we have $G^\Delta - 1$ unknowns. From our previous deliberations concerning the number of unknowns in an equation set compared to the number of equations, we can develop the following conditions:

(1) If $G^\Delta - 1$ is greater than K^{**}, then we have an *underidentified* equation because the number of unknowns is greater than the number of equations. If the number of endogenous variables in the equation minus one is greater than the number of exogenous variables left out of the equation, then we have an *underidentified* equation.

(2) If $G^\Delta - 1$ is exactly the same as K^{**}, we have an *identified* equation because the number of unknowns in the equation is the same as the number of exogenous variables left out of the equation.

(3) If $G^\Delta - 1$ is less than K^{**}, we have an *overidentified* model because we have more equations than unknowns. If the number of exogenous variables left out of the equation is greater than the number of endogenous variables minus one in the equation, we have an *overidentified* model.

This relationship between the number of unknown endogenous variables in an equation and the number of exogenous variables left out of the equation is the *order condition for identifiability*.

Let us return to our triadic arms-race example and use the order condition. We should arrive at the same conclusions as we did with the extended analysis at that time. The structural set was

$$Y_{1t} = \beta_{12}Y_{2t} + \gamma_{10}X_{0t} + u_{1t} \qquad \text{nation A's line}$$

$$Y_{2t} = \beta_{21}Y_{1t} + \gamma_{20}X_{0t} + \gamma_{22}X_{2t} + u_{2t} \qquad \text{nation B's line}$$

$$Y_{3t} = \beta_{31}Y_{1t} + \beta_{32}Y_{2t} + \gamma_{30}X_{0t} + \gamma_{33}X_{3t} + u_{3t} \qquad \text{nation C's line}$$

Let us look first at the line for nation A. The number of endogenous variables in the equation is 2, i.e., Y_{1t} and Y_{2t}. Thus $G^\Delta - 1 = 1$. The number of exogenous variables left out of the equation is also 2, i.e., X_{2t} and X_{3t}. Thus $K^{**} = 2$. Since 2 is greater than 1, we conclude that this equation is overidentified. The same conclusion was obtained in the previous analysis of this equation.

Let us now proceed to B's equation. In this equation there are again two endogenous variables, so that $G^\Delta - 1 = 1$. There is only 1 exogenous variable left out of the equation, i.e., X_{3t}. Thus $K^{**} = 1$. Since $G^\Delta - 1 = K^{**}$, we have an exactly identified equation. This is what we concluded previously.

Finally, let us look at the equation for nation C. In this equation there are three endogenous variables, and thus $G^\Delta - 1 = 2$. The number of exogenous variables left out of the equation is 1, i.e., X_{2t}; thus $K^{**} = 1$. Since $G^\Delta - 1$ is greater than K^{**}, we conclude that this equation is underidentified. This result concurs with our previous one.

The order condition for identifiability provides a rule whereby we can establish the condition of any equation by inspection alone.

However, the order condition for identifiability is only a necessary condition. It is not a sufficient condition because there may not be K^{**} *independent* equations. If the number of independent equations is less than K^{**}, then we may not have enough independent equations for the $G^{\Delta} - 1$ unknowns. Thus we should be concerned with the rank of the matrix $\pi_{\Delta**}$. The rank condition for identifiability is

$$\text{rank}\,(\pi_{\Delta**}) = G^{\Delta} - 1 \tag{12.10}$$

This is *necessary and sufficient* condition for identification.

The rank of $\pi_{\Delta**}$ is obtained by the following process. Rearrange the **B** and Γ matrices into the partitioned form.

$$\mathbf{B} = \left[\begin{array}{c|c} \boldsymbol{\beta}_{\Delta} & \mathbf{0}_{\Delta\Delta} \\ \hline \mathbf{B}_{\Delta} & \mathbf{B}_{\Delta\Delta} \end{array}\right] \quad (12.11) \qquad \Gamma = \left[\begin{array}{c|c} \boldsymbol{\gamma}_{*} & \mathbf{0}_{**} \\ \hline \Gamma_{*} & \Gamma_{**} \end{array}\right] \quad (12.12)$$

where $\boldsymbol{\beta}_{\Delta}$, $\mathbf{0}_{\Delta\Delta}$, $\boldsymbol{\gamma}_{*}$ and $\mathbf{0}_{**}$ are the row vectors.

$\boldsymbol{\beta}_{\Delta}$ is a $(1 \times G^{\Delta})$ row vector

$\mathbf{0}_{\Delta\Delta}$ is a $(1 \times G^{\Delta\Delta})$ row vector

$\boldsymbol{\gamma}_{*}$ is a $(1 \times K^{*})$ row vector

$\mathbf{0}_{**}$ is a $(1 \times K^{**})$ row vector

$\mathbf{B}_{\Delta}$ is a $(G - 1 \times G^{\Delta})$ matrix

$\mathbf{B}_{\Delta\Delta}$ is a $(G - 1 \times G^{\Delta\Delta})$ matrix

Γ_{*} is a $(G - 1 \times K^{*})$ matrix

Γ_{**} is a $(G - 1 \times K^{**})$ matrix

We put the equation we are interested in at the top of the **B** and Γ matrices. We partition each matrix horizontally under the equation. We then rearrange the rows of the matrices so that for the equation under scrutiny all the included variables are to the left and the excluded variables to the right of a vertical partition line. It should be understood that whole columns are moved in this rearrangement, not only the elements in the first row. The rearrangement is shown below.

equation under
consideration $\rightarrow$
$$\mathbf{B} = \left[\begin{array}{c|c} \text{nonzero } \beta\text{'s} & 0\ \beta\text{'s} \\ \hline \mathbf{B}_{\Delta} & \mathbf{B}_{\Delta\Delta} \end{array}\right] \qquad \Gamma = \left[\begin{array}{c|c} \text{nonzero } \gamma\text{'s} & 0\ \gamma\text{'s} \\ \hline \Gamma_{*} & \Gamma_{**} \end{array}\right]$$

We now produce another matrix according to the following:

$$\Delta = [\mathbf{B}_{\Delta\Delta}\Gamma_{**}] \tag{12.13}$$

And it can be shown that:

$$\text{rank}(\pi_{\Delta**}) = \text{rank}(\Delta) - G^{\Delta\Delta} \tag{12.14}$$

We then compare this quantity with the quantity $G^\Delta - 1$. If the rank of $\pi_{\Delta**}$ is equal to $G^\Delta - 1$, then we have satisfied the rank condition. However, if the rank of $\pi_{\Delta**}$ is less than $G^\Delta - 1$, then we have under-identification.

Let us look at this whole process using our arms-race model for the three nations. We arrange the equations (12.1) into the general form:

$$Y_{1t} - \beta_{12}Y_{2t} + 0 \cdot Y_{3t} - \gamma_{10}X_{0t} + 0 \cdot X_{2t} + 0 \cdot X_{3t} = u_{1t}$$

nation A's line

$$-\beta_{21}Y_{1t} + Y_{2t} + 0 \cdot Y_{3t} - \gamma_{20}X_{0t} - \gamma_{22}X_{2t} + 0 \cdot X_{3t} = u_{2t}$$

nation B's line

$$-\beta_{31}Y_{1t} - \beta_{32}Y_{2t} + Y_{3t} - \gamma_{30}X_{0t} + 0 \cdot X_{2t} - \gamma_{33}X_{3t} = u_{3t}$$

nation C's line

Thus

$$\mathbf{B} = \begin{bmatrix} 1 & -\beta_{12} & 0 \\ -\beta_{21} & 1 & 0 \\ -\beta_{31} & -\beta_{32} & 1 \end{bmatrix} \quad \Gamma = \begin{bmatrix} -\gamma_{10} & 0 & 0 \\ -\gamma_{20} & -\gamma_{22} & 0 \\ -\gamma_{30} & 0 & -\gamma_{33} \end{bmatrix}$$

Let us look at nation A's line.

equation under
inspection →

$$\mathbf{B} = \begin{bmatrix} 1 & -\beta_{12} & 0 \\ -\beta_{21} & 1 & 0 \\ -\beta_{31} & -\beta_{32} & 1 \end{bmatrix} \quad \Gamma = \begin{bmatrix} -\gamma_{10} & 0 & 0 \\ -\gamma_{20} & -\gamma_{22} & 0 \\ -\gamma_{30} & 0 & -\gamma_{33} \end{bmatrix}$$

$$\Delta = \begin{bmatrix} 0 & -\gamma_{22} & 0 \\ 1 & 0 & -\gamma_{33} \end{bmatrix}$$

From what we learned in Chapter 6, we see that rank $(\Delta) = 2$.

$$\begin{aligned} \text{rank}(\pi_{\Delta**}) &= \text{rank}(\Delta) - G^{\Delta\Delta} \\ &= \text{rank}(\Delta) - 0 \text{ B's in first equation} \\ &= 2 - 1 = 1 \end{aligned}$$

We know that $G = 2$, i.e., the number of nonzero β's in the first equation.

From (12.14),

$$\text{rank} \, (\pi_{\Delta**}) = G^\Delta - 1$$

i.e.,

$$1 = 1$$

Therefore the first equation is identified according to rank condition. We can look at B's line now and rearrange the rows.

$$
\begin{array}{c}
\text{equation under} \\
\text{inspection} \rightarrow
\end{array}
\mathbf{B} =
\left[
\begin{array}{cc|c}
-\beta_{21} & 1 & 0 \\
\hline
1 & -\beta_{12} & 0 \\
-\beta_{31} & -\beta_{32} & 1
\end{array}
\right]
\quad
\Gamma =
\left[
\begin{array}{cc|c}
\gamma_{20} & -\gamma_{22} & 0 \\
\hline
-\gamma_{10} & 0 & 0 \\
-\gamma_{30} & 0 & -\gamma_{33}
\end{array}
\right]
$$

$$
\Delta =
\begin{bmatrix}
0 & 0 \\
1 & -\gamma_{33}
\end{bmatrix}
$$

and rank $(\Delta) = 1$

$$
\begin{aligned}
\text{rank} \, (\pi_{\Delta**}) &= \text{rank} \, (\Delta) - G^{\Delta\Delta} \\
&= 1 - 1 = 0
\end{aligned}
$$

But we know that $G^\Delta - 1$ in this equation is equal to 1. We conclude, therefore, that this equation does not satisfy the rank criterion for identifiability although we know that it did satisfy the order criterion. This result underlines the importance of carrying out both tests of identifiability.

And finally we look at nation C's arms-level line. Notice how we rearrange the columns of Γ as well as the rows.

$$
\begin{array}{c}
\text{equation under} \\
\text{inspection} \rightarrow
\end{array}
\mathbf{B} =
\left[
\begin{array}{ccc}
-\beta_{31} & -\beta_{32} & 1 \\
\hline
-\beta_{21} & 1 & 0 \\
1 & -\beta_{12} & 0
\end{array}
\right]
\quad
\Gamma =
\left[
\begin{array}{cc|c}
-\gamma_{30} & -\gamma_{33} & 0 \\
\hline
-\gamma_{20} & 0 & -\gamma_{22} \\
-\gamma_{10} & 0 & 0
\end{array}
\right]
$$

$$
\Delta =
\begin{bmatrix}
-\gamma_{22} \\
0
\end{bmatrix}
$$

and

$$\text{rank} \, (\Delta) = 1$$

$$
\begin{aligned}
\text{rank} \, (\pi_{\Delta**}) &= \text{rank} \, (\Delta) - G^{\Delta\Delta} \\
&= 1 - 0 = 1
\end{aligned}
$$

Rank condition test:

$$\text{rank } (\pi_{\Delta**}) = G^\Delta - 1$$
$$1 = 3 - 1 = 2$$

This line is underidentified.

Notice that we have come to slightly different conclusions about the identifiability status of the equations after the rank condition was tested. Whereas with both criteria, lines A and C held the same identification status, line B did not. We can put down table 12.1 concerning identification status.

Table 12.1

	Order condition (necessary)	Rank condition (sufficient)	Identification status
Situation 1.	overidentified	exactly identified	overidentified
Situation 2.	overidentified	underidentified	underidentified
Situation 3.	exactly identified	exactly identified	exactly identified
Situation 4.	exactly identified	underidentified	underidentified
Situation 5.	underidentified	—	underidentified

Obviously, identified and overidentified models allow us to procure the estimates of the structural coefficients, but with underidentified models we are stopped; as it stands, there just is not enough information in the system. What can we do in this situation? First we can come up with some restriction on the structural coefficient for the equation in difficulty. For instance in our arms-race structure, the equation for the C's line is underidentified, but if we, for theoretical reasons, can say that either β_{31}, β_{32}, γ_{30}, or γ_{33} is equal to some quantity, we would have three unknowns and three equations. For instance, if we say that $\gamma_{30} = 0.8$, then the equation set becomes

$$\pi_{30} = \beta_{31}\pi_{10} + \beta_{32}\pi_{20} + 0.80$$
$$\pi_{32} = \beta_{31}\pi_{12} + \beta_{32}\pi_{22}$$
$$\pi_{33} = \beta_{31}\pi_{13} + \beta_{32}\pi_{23} + \gamma_{33}$$

The line becomes identified. But the student should be very careful in doing this. There has to be some reason for suggesting the value of a structural coefficient. It cannot be dreamed up just to overcome the difficulties of an underidentified equation.

A second way to overcome such a difficulty, and I will only mention it briefly here, is to make restrictions about the covariance of the error terms of the structural equations. Since these come into reduced-form equations, we can say that if some of them are independent of each other we are allowed extra equations for our unknowns.

Summarizing this section, we see that there are varying amounts of information in each equation in a structural set. This information is used to obtain estimates of the coefficients in the equation. If there is not enough information, the equation is underidentified. If there is exactly the right amount of information, the equation is identified. If there is too much information, then the equation is overidentified.

12.3   Single-equation estimation

We have seen that there are two difficulties involved when we deal with simultaneous equations. Specifically, estimates of the coefficients in some equations may not exist, and in those equations where they do exist, they are generally not consistent. If the estimates do not exist (the under-identified situation), there is nothing that can really be done by the politometrician unless further information is forthcoming. If the estimates do exist (the identified and overidentified situation), we require techniques for obtaining them so that they are consistent.

There are a number of different methods for obtaining consistent estimates. Some of them involve estimating each equation individually. This is called *single equation* estimation or, alternatively, *limited information* estimation since it does not require the use of all the data on all the variables in the structural system. Others involve the estimation of the whole structural set at one time. This is called *system estimation* or *full information estimation.*

We will deal with single-equation estimation only. As we shall see, such procedures cope quite easily with all of the situations we are liable to confront. The emphasis on single-equation estimation rather than multi-equation or system estimation can be argued for two reasons. First, in the system estimation, error from one equation slops over into the estimates in other equations. For example, if we have a variable in one equation which has a large error component, this error component will affect the estimates in other equations whether or not the offending variable is included in those equations. Secondly, misspecification in one equation will produce similar unwanted effects in other equations if system estimation is used.

Before we look at a single equation estimation process let us again consider the inconsistency problem. The basic problem is that inconsistent estimates are obtained because endogenous variables acting as explanatory variables in the equation are correlated with the error term. For example, in our overidentified arms-race equation for nation A, we know that Y_{2t} is correlated with u_{1t}. We know this because if we were to solve the following two equations for Y_{2t}

$$Y_{1t} = \beta_{12}Y_{2t} + \gamma_{10} + u_{1t}$$

$$Y_{2t} = \beta_{21}Y_{1t} + \gamma_{20} + \gamma_{22}X_{2t} + u_{2t}$$

we would obtain

$$Y_{2t} = \left[\frac{\beta_{21}\gamma_{10} + \gamma_{20}}{1 - \beta_{21}\beta_{12}}\right]X_{0t} + \left[\frac{\gamma_{22}}{1 - \beta_{21}\beta_{12}}\right]X_{2t} + \left[\frac{\beta_{21}u_{1t} + u_{2t}}{1 - \beta_{21}\beta_{12}}\right]$$

We see that Y_{2t} is dependent upon u_{1t}. We need some technique for getting rid of the effects of u_{1t} on Y_{2t}. That is, we need to purge Y_{2t} of u_{1t}. Let us consider how this can best be achieved.

In the reduced-form equation for Y_{2t} we have

$$Y_{2t} = \pi_{20} + \pi_{22}X_{2t} + \pi_{23}X_{3t} + v_{2t}$$

And from above we see that

$$v_{2t} = \frac{\beta_{21}u_{1t} + u_{2t}}{1 - \beta_{21}\beta_{12}}$$

If we can purge v_{2t} from Y_{2t} we will also achieve our aim of purging u_{1t} from Y_{2t}. We need to produce the relationship $Y_{2t} - v_{2t}$. But from our deliberations in Part One we know that

$$Y_{2t} - v_{2t} = \hat{Y}_{2t}$$

Therefore we know that $\hat{Y}_{2t}$ will have no component of u_{1t} in it. If we use $\hat{Y}_{2t}$ in our equation so that

$$Y_{1t} = \beta_{12}\hat{Y}_{2t} + \gamma_{10} + u_{1t}$$

our inconsistency problem will have been removed.

This suggests the two stages of two-stage–least-squares regression (2SLS):

> Stage 1: Obtain estimates of all endogenous variables acting as explanatory variables in the overidentified or exactly identified equations under consideration. To accomplish this, use the reduced-form equations.
>
> Stage 2: Replace all of the endogenous variables acting as explanatory variables in the overidentified or exactly identified equations by estimates of them obtained from the previous stage. Carry out ordinary least-squares regression on these data.

The estimates obtained in this manner are consistent estimates of the structural coefficients, and the variances of the estimates are also consistent.

Let us perform this whole process for our overidentified equation for the arms level of nation A. The equation here is

$$Y_{1t} = \beta_{12}Y_{2t} + \gamma_{10}X_{0t} + u_{1t}$$

An imaginary data set might be as shown in table 12.2.

Table 12.2

	Y_{1t}	Y_{2t}	Y_{3t}	X_{2t}	X_{3t}		Y_{1t}	Y_{2t}	Y_{3t}	X_{2t}	X_{3t}
$t = 1$	7	4	3	5	1	$t = 14$	2	7	7	8	6
$t = 2$	1	2	8	4	9	$t = 15$	2	8	6	5	9
$t = 3$	6	5	9	2	1	$t = 16$	7	1	6	9	5
$t = 4$	4	2	3	5	6	$t = 17$	1	3	3	7	4
$t = 5$	1	3	5	4	2	$t = 18$	5	2	7	4	6
$t = 6$	4	4	3	4	2	$t = 19$	2	1	3	2	1
$t = 7$	7	2	0	2	8	$t = 20$	9	5	1	3	1
$t = 8$	0	5	8	1	0	$t = 21$	5	8	9	0	1
$t = 9$	5	3	5	7	4	$t = 22$	7	7	9	9	7
$t = 10$	3	1	9	3	7	$t = 23$	1	6	1	6	9
$t = 11$	3	8	8	7	3	$t = 24$	3	8	3	5	4
$t = 12$	5	7	2	6	5	$t = 25$	9	5	2	4	1
$t = 13$	7	1	2	3	4						

To purge Y_{2t} of u_{1t} we need to obtain the data set $\hat{Y}_{2t}$. To do this we regress the reduced form equation.

$$Y_{2t} = \pi_{20} + \pi_{22}X_{2t} + \pi_{23}X_{3t} + v_t$$

Using the data, we obtain

$$\hat{Y}_{2t} = 3.7880 + 0.23052X_{2t} + (-0.12463)X_{3t}$$

This produced the $\hat{Y}_{2t}$ data column vector,

$$\hat{Y}_{2t} = \begin{bmatrix} 4.816 \\ 3.588 \\ 4.124 \\ 4.193 \\ 4.461 \\ 4.461 \\ 3.252 \\ 4.019 \\ 4.903 \\ 3.607 \\ 5.028 \\ 4.548 \\ 3.981 \\ 4.884 \\ 3.819 \\ 5.240 \\ 4.903 \\ 3.962 \\ 4.124 \\ 4.355 \\ 3.663 \\ 4.990 \\ 4.049 \\ 4.442 \\ 4.585 \end{bmatrix}$$

We only carry out this procedure for Y_{2t} because this is the only endogenous variable acting as an explanatory variable in the equation. This ends the first stage.

The second stage involves ordinary least-squares regression of the equation

$$Y_{1t} = \beta_{12}\hat{Y}_{2t} + \gamma_{10} + u_{1t}$$

In this manner we obtain the estimating line

$$\hat{Y}_{1t} = 0.77984\hat{Y}_{2t} + 0.87117$$

The estimates in this equation are consistent. The variances for these estimates are calculated in the same way as described in part two of the text, using the $(\mathbf{X'X})^{-1}$ matrix and the residual variance $\hat{\sigma}^2$. This provides a variance of 1.092 for β_{12} and 20.5 for the intercept term. Thus the whole process is achieved using two runs of OLS on selected data. And although we used an overidentified equation for our example, we could carry out precisely the same procedure on an exactly identified equation.

12.4   Other structural models

Finally we should discuss other types of structural models. First we shall look at block recursive models and then look at equations which are seemingly unrelated.

12.4.1   Block recursive models

The recursive model is one variation of the structural model. In this model the regression slope matrix will be

$$\mathbf{B} = \begin{bmatrix} \beta_{11} & 0 & 0 & \cdots & 0 \\ \beta_{21} & \beta_{22} & 0 & & 0 \\ \beta_{31} & \beta_{32} & \beta_{33} & & \cdot \\ \cdot & \cdot & \cdot & & \cdot \\ \beta_{G1} & \beta_{G2} & \cdot & \cdots & \beta_{GG} \end{bmatrix}$$

In this set we have G equations. Notice that the matrix is triangular. The student will remember that this allowed us to consider variables higher in the equation set as exogenous to those lower in the set, despite the higher variables acting as endogenous variables in the higher equations. One can also achieve something similar to this with blocks of variables rather than individual variables. In this setup it is blocks of variables that form the triangular matrix. In the matrix below, the elements of the matrix are denoted by numbered suffixes.

$$\mathbf{B} = \begin{bmatrix} \mathbf{B}_{11} & \mathbf{0} & \mathbf{0} & \cdots & \mathbf{0} \\ \mathbf{B}_{21} & \mathbf{B}_{22} & & & \cdot \\ \mathbf{B}_{31} & \mathbf{B}_{32} & & & \cdot \\ \cdot & \cdot & & & \cdot \\ \cdot & \cdot & \cdots & \cdots & \mathbf{0} \\ \mathbf{B}_{P1} & \mathbf{B}_{P2} & \cdots & \cdots & \mathbf{B}_{PP} \end{bmatrix}$$

In this matrix we assume that there are P blocks of equations.

An important gain is made if the researcher can rearrange the equation system so that it conforms to the above configuration. As with the variable recursive situation, the various blocks higher in the matrix can be seen as exogenous to those lower in the matrix, as long as error terms from different blocks are not related to one another. Thus blocks of variables, which within themselves may not be recursive, can be solved entirely without reference to those lower in the structure. This can help in situations where equations lower in a series of equations are under-identified. If we can achieve a block recursive configuration, then some of the seemingly endogenous variables can be considered as exogenous, reducing the number of endogenous variables to exogenous variables—the criterion for identification.

Let us illustrate with the Gurr and Duvall example on civil violence. In Chapter 11 we saw that the **B** matrix for their model had the formation

$$\mathbf{B} = \begin{bmatrix} 1 & -\beta_{12} & 0 & 0 & 0 & 0 & 0 \\ -\beta_{21} & 1 & -\beta_{23} & 0 & 0 & 0 & 0 \\ 0 & -\beta_{32} & 1 & 0 & 0 & 0 & 0 \\ 0 & 0 & -\beta_{43} & 1 & -\beta_{45} & 0 & 0 \\ 0 & -\beta_{52} & 0 & 0 & 1 & -\beta_{56} & 0 \\ -\beta_{61} & 0 & 0 & 0 & -\beta_{65} & 1 & -\beta_{67} \\ 0 & 0 & 0 & -Y_{3t} & 0 & 0 & 1 \end{bmatrix}$$

This matrix can be represented by

$$\mathbf{B} = \begin{bmatrix} \mathbf{B}_{11} & \mathbf{0} \\ \mathbf{B}_{21} & \mathbf{B}_{22} \end{bmatrix}$$

where

$$
\mathbf{B}_{11} = \begin{bmatrix} 1 & -\beta_{12} & 0 \\ -\beta_{21} & 1 & -\beta_{23} \\ 0 & -\beta_{32} & 1 \end{bmatrix} \quad \mathbf{B}_{21} = \begin{bmatrix} 0 & 0 & -\beta_{43} \\ 0 & -\beta_{52} & 0 \\ -\beta_{61} & 0 & 0 \\ 0 & 0 & 0 \end{bmatrix}
$$

$$
\mathbf{B}_{22} = \begin{bmatrix} 1 & -\beta_{45} & 0 & 0 \\ 0 & 1 & -\beta_{56} & 0 \\ 0 & -\beta_{65} & 1 & -\beta_{67} \\ -Y_{3t} & 0 & 0 & 1 \end{bmatrix}
$$

This is block recursive. The block of coefficients $\mathbf{B}_{11}$ are not recursive within themselves and can be estimated using 2SLS. However, block $\mathbf{B}_{11}$ is recursive to both $\mathbf{B}_{21}$ and $\mathbf{B}_{22}$. Thus the variables in $\mathbf{B}_{11}$ can be seen as exogenous in $\mathbf{B}_{21}$ and $\mathbf{B}_{22}$ as long as we can assume that error terms between blocks are unrelated, although error terms within blocks may still be related. Had one of the equations in $\mathbf{B}_{21}$ or $\mathbf{B}_{22}$ been underidentified (obviously this is determined by the exogenous variables which we have not considered here) then the change of the variables in $\mathbf{B}_{11}$ from endogenous to exogenous by this block recursive maneuver may very well have helped us out of the underidentified problem. Gurr and Duvall actually used this procedure in their analysis, although they had no particular problems with identification.

12.4.2 Seemingly unrelated equations

Consider the following $\mathbf{B}$ matrix:

$$
\mathbf{B} = \begin{bmatrix} \beta_{11} & 0 & \cdots & \cdots & 0 \\ 0 & \beta_{22} & & & \cdot \\ 0 & 0 & \beta_{33} & & \cdot \\ \cdot & & & & \cdot \\ 0 & \cdots & \cdots & \cdots & \beta_{GG} \end{bmatrix}
$$

Since the researcher will endeavor to have the explained variable on the left of the equal sign in the equations, the set of equations may appear to be unrelated rather than simultaneous. This however assumes that the error terms in each equation are unrelated. Unless this can be assumed, we have to consider the set of equations as simultaneous and confront the problems of identification and inconsistency. Unless we can assume that the error terms from each equation are unrelated, we have to carry out estimation using indirect or two-stage least-squares.

We can extend this argument to consider blocks of variables. In the matrix below, each block appears to be unrelated to any other block of equations.

$$
B = \begin{bmatrix}
B_{11} & 0 & \cdots & \cdots & 0 \\
0 & B_{22} & \cdots & \cdots & \cdot \\
\cdot & \cdot & \cdots & \cdots & \cdot \\
0 & \cdots & \cdots & \cdots & B_{PP}
\end{bmatrix}
$$

If, indeed, the error terms from any one block are unrelated to those of another block, we can treat each block of equations as a separate entity for all estimation purposes. But if we are unable to assume unrelated error terms we have P seemingly unrelated blocks of variables. Since they are related through the error terms, we have to take this into account when estimating the coefficients.

Further readings

Substantive
Choucri, N., and R. C. North. 1975. *Nations in Conflict.* San Francisco: Freeman.
Gurr, T. R., and R. Duvall. 1973. "Civil Conflict in the 1960's: A Reciprocal Theoretical System with Parameter Estimates." *Comparative Political Studies* 6: 135.
Hibbs, D. A. 1973. *Mass Political Violence: A Cross-National Causal Analysis.* New York: Wiley.

Statistical
Goldberger: pp. 329–38.
Johnston: pp. 380–84.
Kmenta: pp. 550–72.
Wonnacott and Wonnacott: pp. 190–92, 358–64.

Index

Added observations test, 175–77
Added variables test, 172–75
Analysis of variance table, 166, 168
Autocorrelation of error terms, 45, 84, 243
  discovery, 84–89
  remedial action, 103–10

Binary (dummy) variables, 187–200
Blalock, H., 219, 220, 224, 226, 233, 235

Campbell, D. S., 195
Causal inference, 224–37
  discrimination problem, 227
Causation, 5–9
Coefficient of determination (R^2), 149–53
  corrected coefficient of determination
  ($\bar{R}^2$), 153–54
  matrix form, 154
Confidence intervals
  explained variable, 73–76, 178–80
  regression intercept, 72–73, 177–78
  regression slope, 70–72, 177–78
Correlation
  coefficient, 52–58
  Fisher transformation, 67
  versus regression, 6–7

Degeneracy, 46, 62
Degrees-of-freedom problems, 155–56
Discriminant functions, 197–98
Draper N. R., 213

Durbin, J., 107, 108
  method for ρ, 107
  Durbin-Watson statistic, 85–89
  Durbin-Watson statistic for ρ, 107, 108, 110
Duvall, R., 7–9, 278–79

Error terms
  in equations, 30–34
  nonlinear estimation, 205–6
  violation of assumptions, 93–110
Estimates
  Bias, 183–84
  BLUE, 20, 52, 93
  characteristics, 14–20
  consistent, 18–19, 25, 52
  efficient, 17–18
  relative efficiency, 17
  unbiased, 16–17, 24, 52

Generalized differences, 105–6
Goldberger A., 19, 53, 164
Gurr, T. R., 4–9, 240, 244, 278–79

Heteroscedasticity, 44, 82, 94 ff, 243
  caused by one explanatory variable, 95–100
  caused by more than one explanatory variable, 100–102
  discovery, 82–84
Homoscedasticity, 44

282 Index

Identification, 6
  identified, 261, 264, 265, 268
  order condition, 268–73
  overidentified, 261, 263, 265, 268
  problem of, 239, 251–59, 260–73
  rank condition, 269–73
  underidentified, 261, 264, 267

Johnston, J., 158

Kmenta, J., 19
Kort, F., 197–200

Least-squares regression
  assumptions, 43–46, 139–40
  criterion, 40
  multiple regression, 138
  normal equations, 47–48, 140–41
  violation of assumption, 92–110
Logit analysis, 200

Malinvaud E., 52
Matrix algebra, 118–37
  basic operations, 120–27
  definition, 118
  determinants, 127–31
  matrix inversion, 131–35
  rank, 135–37
Models
  deterministic, 30
  estimator, 37–38
  parametric, 10
  statistical, 36–38
  stochastic, 30
  theoretical, 36–38
Multicollinearity, 187, 156–59, 168, 202, 210

Path Analysis, 229–237
Populations, hypothetical, 26–27
Prediction, 76–78, 180–82
Probit analysis, 200

Recursive models, 6, 219
  block, 277
Reduced form equations, 246–50, 263, 264
  coefficients, 263–265

Regression models
  lagged variables, 208–11
  linear, 36–52
  logarithmic transformation, 203–25
  missing observations, 213–14
  misspecification, 182–84
  nonlinear, 200–205
  polynomial, 201–2
  reciprocal, 201–3
  restricted coefficient, 206–8
  steady growth, 203–4
  stepwise procedures, 212–18
  time series problems, 211–12
Residuals, 37–38, 144–46
Richardson, L. F., 247

Samples, 12–13
Significance testing
  correlation coefficient, 66
  regression intercept, 69–70, 160–65
  regression slope, 64–66, 160–65
Simon, H., 213
Stochastic explanatory variables, 111
Structural systems, 242
  coefficients, 242, 260, 264, 272–73
  inconsistency problem, 239, 250–51, 274–75
  models, 239 ff.
System estimation, 273–74

Testing hypotheses
  general, 60–64
  whole regression lines, 165
Two-stage least-squares (2SLS), 275–77

Variables
  endogenous, 240–66
  exogenous, 240–48
  predetermined, 240
Variance of regression estimates, 61–64, 146–48, 178

Weighted least-squares, (WLS), 95
  generalized, 102–3
Wright, S., 219, 220, 228

Zinnes, D., 209